Gorgeous BEASTS

ANIMALIBUS VOL. 2
OF ANIMALS AND CULTURES

Books in the Animalibus series share a fascination with the status and the role of animals in human life. Crossing the humanities and the social sciences to include work in history, anthropology, social and cultural geography, environmental studies, and literary and art criticism, these books ask what thinking about nonhuman animals can teach us about human cultures, about what it means to be human, and about how that meaning might shift across times and places.

Other titles in the series:

Rachel Poliquin *The Breathless Zoo: Taxidermy and the Culture of Longing*

Gorgeous BEASTS

ANIMAL BODIES IN HISTORICAL PERSPECTIVE

EDITED BY
Joan B. Landes, Paula Young Lee,
& Paul Youngquist

THE PENNSYLVANIA STATE UNIVERSITY PRESS
UNIVERSITY PARK, PENNSYLVANIA

Funding for this project was provided by the Penn State Institute for Arts and Humanities.

Library of Congress Cataloging-in-Publication Data
Gorgeous beasts : animal bodies in historical perspective / edited by Joan B. Landes, Paula Young Lee, and Paul Youngquist.
p. cm.
Includes bibliographical references and index.
Summary: "A collection of essays examining the place of animals in history and culture and their influence on life and art, from the Renaissance to the present"—Provided by publisher.
ISBN 978-0-271-05402-5 (pbk : alk. paper)
1. Animals and civilization—History.
2. Human-animal relationships—History.
3. Animals and history.
I. Landes, Joan B., 1946– .
II. Lee, Paula Young.
III. Youngquist, Paul.

QL85.G67 2012
590—dc23
2012009331

Printed in the United States of America
Published by
The Pennsylvania State University Press,
University Park, PA 16802-1003

The Pennsylvania State University Press is a member of the Association of American University Presses.

It is the policy of The Pennsylvania State University Press to use acid-free paper. Publications on uncoated stock satisfy the minimum requirements of American National Standard for Information Sciences—Permanence of Paper for Printed Library Material, ANSI Z39.48–1992.

Designed by Regina Starace

contents

illustrations

COLOR PLATES *(following page 178)*

1 Mark Dion, *Les nécrophores–L'enterrement (Hommage à Jean-Henri Fabre),* 1997. Synthetic moleskin, rope, synthetic necrophore beetles, jute rope, eye hooks, books. Installation view from the exhibition Mark Dion: Microcosmographia, South London Gallery, London, 2006. Collection Antoine de Galbert, Paris. Photo courtesy of the artist and Tanya Bonakdar Gallery, New York.

2 Mark Dion, *Park: Mobile Wilderness Unit,* 2001. Construction trailer, taxidermic European bison, painted backdrop, dirt, leaves, stones, artificial fern and fungi, 290 x 170 x 380 cm. Photo courtesy of the artist and Tanya Bonakdar Gallery, New York.

3 Mark Dion, *Polar Bears and Toucans (from Amazonas to Svalbard),* 1991. Stuffed toy polar bear, Sony sport cassette player, cassette recorded in Venezuela/Amazonas territory, shipping crate, electrical cord, tar, wash tub, 231 x 112 x 75 cm. Collection DuMont Schütte. Photo courtesy of the artist and Tanya Bonakdar Gallery, New York.

4 Mark Dion, *Landfill,* 1999–2000. Mixed media, 71.5 x 147.5 x 64 in. Collection of Museum of Contemporary Art, San Diego, California. Courtesy the artist and Tanya Bonakdar Gallery, New York. Photo: Pablo Mason.

5 Mark Dion, *Survival of the Cutest (Who Gets on the Ark?),* from the series *Wheelbarrows of Progress,* with William Schefferine, 1990. Toy stuffed animals, white enamel on red steel, wood and rubber wheelbarrow, 63.5 x 68.5 x 141 cm. Photo courtesy of the artist and Tanya Bonakdar Gallery, New York.

6 Mark Dion, *Ichthyosaur,* 2003. Mixed media. Photo courtesy of the artist and Tanya Bonakdar Gallery, New York.

7 Mark Dion, *The Delirium of Alfred Russel Wallace,* 1994 (detail). Installation, mixed media. Photo courtesy of the artist and Tanya Bonakdar Gallery, New York.

8 Mark Dion, *Extinction Series: Black Rhino Head,* 1989 (detail). Wooden crates, stenciled lettering, color photographs, rhino head, wood chips, map of Africa. Photo courtesy of the artist and Tanya Bonakdar Gallery, New York.

9 Mark Dion, *Killers Killed,* 1994–2007. Tree, taxidermic animals, tar, galvanized aluminum, foam, paint, 128 x 65 x 60 in. Photo courtesy of the artist and Tanya Bonakdar Gallery, New York.

10 Mark Dion, *Grotto of the Sleeping Bear—Revisited,* 1997. Synthetic bear, plastic, bubble wrap, wood crate, sawhorses. Photo courtesy of the artist and Tanya Bonakdar Gallery, New York.

FIGURES

acknowledgments

The editors thank the Penn State Institute for the Arts and Humanities, Dean Susan Welch, and the College of Liberal Arts for support of this publication, and the members of the Visualizing Animals Interdisciplinary Project and the Finding Animals Conference, from which the book developed.

We are deeply indebted to Mark Dion for his generous contribution to this project. Emily Ruotolo, assistant director of the Tanya Bonakdar Gallery, New York, Fabienne Leclerc and Antoine Laurent of In Situ/Fabienne Leclerc Gallery, Paris, and Matthias Bildstein and Marie Duhnkrack, of Georg Kargl Fine Arts, Vienna, all offered invaluable assistance in identifying and providing photographs of Mark Dion's works. Similarly, we greatly appreciate photographer Amy Stein's contribution to this endeavor.

We are grateful to the many archives, libraries, and galleries that provided images for this book, including Sandy Stelts, curator of rare books and manuscripts, Pennsylvania State University Libraries, and Carol Togneri, chief curator at the Norton Simon Museum in Pasadena, California. It has been a pleasure to work with Kendra Boileau, the editor-in-chief of Pennsylvania State University Press, and editorial assistant Stephanie Lang. We are pleased to appear in the Press's Animalibus series and warmly acknowledge the support of general editors Nigel Rothfels and Garry Marvin.

Introduction

JOAN B. LANDES, PAULA YOUNG LEE,
AND PAUL YOUNGQUIST

Lately there have been foxes. Outside the office window appeared the agouti shape of a lithe interloper. She basked in the winter sun, half-asleep on a stone seat in an adjacent amphitheater, comfortable and incongruous. Later, in London, the wind a hard slap to the face, she appeared again: fluffed and taut, her eyes flashing as she slid through shadows around trash bags, into the dusk. Who is this furred phantasm with her wily flesh?

To encounter an animal, especially unexpectedly, is to wonder, "what is this beast *doing* here?" It's a question for which there are surprisingly few answers. Perhaps this fox is simply *living,* persisting in her vitality, *doing her thing.* Why she should do it between a Cineplex and a tube stop is perplexing, but there you have it. She persists, welcome or not. This fox comes to trouble comfortable assumptions about human privilege and animal obeisance. She lives beyond these distinctions, much like the fox captured in photographer Amy Stein's image *Passage* in her *Domesticated* series (fig. 1). Seated motionless atop the gravel-graded opening of a drainpipe, seemingly suspended between nature and culture, she evokes our paradoxical relationship to the "wild": our desire to be part of nature while simultaneously taming it.

Gorgeous Beasts: Animal Bodies in Historical Perspective names a desire as much as a discovery. Animals realize a life that exceeds the small circle of our so-called humanity, a full and feral life irreducible to reason and its pale twin, propriety. While incessantly alluring to humans, indeed necessary to human

1 Amy Stein, *Passage,* 2008. From Stein, *Domesticated.* Portland: Photolucida/Critical Mass, 2008. Photograph © Amy Stein.

existence, the proximity of animals to humans is a source of both their suffering and our delight. The title foregrounds the doubleness of human relations to animals, that is, the contradictory connections between human and nonhuman animals in different social contexts and places, ranging from early modern to contemporary times. The essays in this collection find animals abundant and audacious: horrific, fierce, tender, or vulnerable. Animals occasion new emotions. They attract and repel. They seduce, while too often becoming objects of unacknowledged violence. They dismantle old beliefs and also challenge humans to devise new ways of living in concert with and among easily overlooked or undervalued species. The book's contributors address various ways in which humans and animals are linked within specific sociopolitical and intellectual contexts, including aristocratic and capitalist class structures, colonial trade and racial domination, liberal theories of rights and modern science. They call attention to the paradoxical ways in which animals have been assigned to the categories "wild" or "natural" and have been consumed and appropriated for different human purposes. These essays also reveal that animal-human relations are shaped by animals' responses to, sympathy for, and interest in their human partners. *Gorgeous Beasts* draws generously upon visual as well as textual sources, integrating artistic with analytical, scientific, or literary responses to animal bodies, and showing how vision and curiosity play a significant role in human responses to animal beauty.

Picture a fox. In a simple sense, this is an act of imagination. A fox, or an image of one, appears in your head, and it represents real-life foxes. The image "fox" (whiskers, teeth, tawny fur, and bushy tail) stands in for the elusive, real, and absent thing. So an imagined picture of a fox is like a fox, only less so—the next-best thing to having one of your own, better, maybe, since foxes are not easy to domesticate. An imagined fox is cute, docile, clean: fox lite.

This is where things get tricky.[1] Foxes may be hard to domesticate, but they come home in pictures. Representations domesticate them, turn them from living beings into signs of life. An image of a fox is a kind of compromise between its life and ours, a mingling of traces that produces a hybrid: neither quite fox nor quite human but something in between. This hybrid is a sign. It bears a resemblance to a fox. But it shows too the active touch of the imagination that produced it. One might ask of any given image of a fox, *which fox does it signify?* All foxes? Just a few? One particular fox in a Soho dumpster or a photograph by Amy Stein?[2]

The work of the artist Mark Dion helps one see how much is involved in seeing animals. His installations and assemblages show how encounters with them take place on many registers at once. Animals are never just *there* to be seen, felt, or known. History situates them. Culture appropriates them (fig. 2). Science defines them in one way, affection in another. Dion's work reveals how much work it takes to see a fox, how much baggage one brings to any encounter with animal bodies. An installation titled *The Delirium of Alfred Russel Wallace,* originally staged in Vienna in 1994, illustrates the point with intelligence and wit (fig. 3). At its center is a fox, stuffed and lying on its back in a hammock (color plate 7). Shrouded in mosquito netting, the hammock hangs between two trees, one living and one dead. Flanking each is an old steamer trunk surrounded by the tools of the naturalist's trade: butterfly nets, specimen boxes, binoculars, maps, books, a shotgun, a bottle, and more. The scene resembles the campsite in Malaysia of the great explorer, naturalist, and co-discoverer of evolution, Alfred Russel Wallace, with one key difference: Wallace *is* the fox. An audible recording of Wallace reflecting on his life in the field completes the identification. Dion forces his viewer to question human encounters with animal bodies by substituting an animal body for that of a historically particular human.

2 Mark Dion, *Scala Naturae,* 1994. Stepped plinth, artifacts, specimens, taxidermic animals, bust, 238 x 100 x 297 cm.

To see the body of a fox here involves perceiving it in human terms. A panoply of discourses, technologies, and customs make it visible. Natural history frames the whole encounter, which is to say that "nature" is a historical invention. The scientific bias of that invention appears in all the gear for measuring and recording its details. But other kinds of tools are involved, too: a pot for cooking, a bottle for drinking, a lantern for driving back the dark. The animal body at the center of Dion's installation appears as the object not so much of science as of the various histories and habits that sustain science as a form of knowledge. History unfurls a British imperialist backdrop to the great naturalist reposing in Malaysia. Culture packs the scene with European accoutrements. That fox in the hammock is the creature of a whole arsenal of tools and beliefs.

To encounter this fox is to acquire an awareness of the complex historical and cultural heritage that enables perception. Dion's soundtrack doubles that awareness, as Wallace confesses his own inescapable bias: "I will often set foot

3 Mark Dion, *The Delirium of Alfred Russel Wallace,* 1994. Installation, mixed media, approx. 320 x 295 x 320 cm. Photo courtesy of the artist and Tanya Bonakdar Gallery.

on a land which no European eye has beheld, or see animals unknown to our world, or be treated to customs completely alien in origin; yet none of it stands in the foreground of my mind. My thoughts may be occupied by visions of the English countryside, or filled with fragments of an absurd and detestable children's song, running a course in my head. London Bridge is falling down."[3] Such is the delirium of this naturalist in the field: part exotic sensation, part national identity, part foolish memory—and part fox. The spectacles resting on its black snout bring its stuffed body into focus as a complex object, composed as much of taxonomy, dream, and desire as of fur and flesh. Dion's crowning irony is the impossibility of the perception his installation nevertheless documents. As the father of bio-geography and the author of *Geographical Distribution of Animals,* Wallace knew what his scientific treatise established: in Malaysia there are no foxes. Yet there one sleeps, conjured by science, culture, and delirium. The animal bodies encountered in *Gorgeous Beasts* share these contingent origins.

Marks of the Beast

Encountering animals turns out to be a much more complicated process than it looks. Representing them in signs requires a lot of stage setting before one can say with assurance, "oh, *that* fox," "oh, *that* cat." For the sake of simplicity, let's say it involves at least two registers of perception, one erotic and the other historical. An old adage is worth pondering: people see what they want to see. Seeing involves desire.[4] The (misogynist) popular history of the word "fox" makes the point obvious. Images, words, signs get tinged by the desire that invokes them. To focus perception on *this* particular object rather than that is to mark it as wanted. Seeing animals means marking them, making and remaking them in the image of desire. The other register of perception is historical. Any given image comes trailing a long history of representations that enables us to identify it with a fox or a cat. One important implication of this observation is that *seeing is an act of history.*[5] Seeing animals means seeing them *through* a long, complex, invisible history of representations that name them, locate them, and value them, making it possible to exclaim, "*that's* a fox!"

Animals inhabit environments altered by contact with human societies, just as human relationships are mediated by animals co-present in their various locales.[6] Consider another familiar (domesticated) beast, the horse: an animal that has cohabitated with humans for millennia, playing a prominent role in human society. In agriculture, horses pulled plows, provided fertilizer and even food—as any visitor to a French *boucherie chevaline,* specializing in horse meat, can attest. Horses facilitated long-distance trade and played a decisive role in military conquest. Their hides provided leather to human populations. Given that we commonly think of horses as part of either a natural or a rural landscape, the horse's role in modern urban life and industry can come as a surprise. The authors of *The Horse in the City: Living Machines in the Nineteenth Century* present the nineteenth-century city as the climax of human exploitation of horse power, crediting the horse with helping to build that age's giant, wealth-generating metropoles. Yet this is not a simple story of animal as object or victim, human as exploiter. Surprisingly, horses also benefited from the new human ecology: "Their populations boomed, and the urban horse, although probably working harder than his rural counterpart, was undoubtedly better fed, better housed, and protected from cruelty. To the extent that it can be determined, the urban horse was also larger and longer lived than were farm animals. Thus the relationship was symbiotic—horses could not have survived as a species without human intervention, and dense human populations frequently relied on horses. . . . The European horse survived because it found an ecological niche as a partner for

4 Amy Stein, *New Homes,* 2008. From Stein, *Domesticated.* Portland: Photolucida/Critical Mass, 2008. Photograph © Amy Stein.

humans. In a sense this was co-evolution, not domination."[7] Co-survivor, co-evolver, and (no matter how lopsided) companion animal: the horse's integration into the story of the nineteenth century's emergent metropolis demands a more complex account than everyday assumptions allow, particularly when those assumptions require us to draw lines between what is domestic and what is wild, what is cultural and what is natural.[8]

Amy Stein's *Domesticated* series brings these issues into focus. In an image titled *New Homes,* two bobcats range unexpectedly into human territory (fig. 4). One sits on a poured concrete foundation, sheltered by a skeleton frame made of wood. Another sits atop a wrapped package of lumber lying next to a pile of exposed planks. Predators at rest, these bobcats are seductive diplomats of danger, night watchers in daylight waiting for prey: rabbits hiding in tall weeds. Arrestingly lucid, the photograph resembles the kind of "Otherwhere" images published in *National Geographic,* which often highlight quixotic creatures

5 Amy Stein, *Dead End Street,* 2007. From Stein, *Domesticated.* Portland: Photolucida/Critical Mass, 2008. Photograph © Amy Stein.

wandering on the socioeconomic margins. The photographer is skilled. The viewer may feel admiration for the quicksilver reflex of the shutterbug's eye. There seem to be luck, patience, and a story here, an allegory regarding the largesse of carnivores. The apparent poetry of the scene suggests abandonment, as if humans have departed forever, letting other predators take their place in nature's hierarchy.

The story is all too familiar. The everyday experience of animals comes filtered through these sorts of waiting-room images. The brilliance of Stein's photograph blooms in its familiarity, much like the famous bunch of grapes painted so skillfully by Zeuxis that birds tried to peck at them. According to the Roman historian Pliny, Zeuxis turned to his rival, Parrhasius, and asked him to reveal his own masterpiece. "Pull back the curtain!" Parrhasius ordered. Certain of his superiority, Zeuxis reached forward. Only then did he realize that there was nothing to pull back: the curtain was a painted image! Conceding defeat, Zeuxis muttered that his own work had fooled animals, but his rival had fooled another artist. Stein's photograph adds the latest wrinkle to this famous challenge. This isn't a lucky snapshot in the genre of travel photography but a staged tableau that exposes norms informing animal-human relationships. We are Zeuxis gazing blindly at Parrhasius's curtain. The bobcats aren't bobcats. They are taxidermic specimens, closer to stuffed handbags than to cunning killers. Photographically alive, they are technically dead. To see them as objects on display transforms the "homes" they haunt into mausoleums under construction. This purring paradox of the animal image stalks our backyards.

At first glance, Stein's photograph *Dead End Street* (fig. 5) seems to be about nothing in particular, in striking contrast to the other images in *Domesticated,*[9] which abound with fawns, bears, canaries, foxes, bobcats, and other charismatic creatures. This image of a dead-end street, however, reveals something brutal in its banality. The natural scene appears wholly under human control. Manicured and monotonously green, the surrounding forest is as man-made as the human homes lining the paved streets, making us understand that images of nature are, in fact, unnatural. Posed on the edge of irony, Stein's fox in a drainpipe (fig. 1), one of numerous stuffed specimens, forces us to confront the limits of human understanding, specifically that of sight as the sense tied most closely to reason. Seeing and knowing are worlds apart. What do we really know about those houses? Perhaps there are people inside them. Maybe animals live in those woods and wild birds fly overhead. We do not know. We cannot see them. That uncertainty condemns us to know animals primarily through images.

"Ambiguity," the art historian Ernst Gombrich declared, is "clearly the key to the whole problem of image reading."[10] The postmodern position exploits this ambiguity by stressing a fluid morphology in a Darwinian universe where neither human nor animal assumes a clear priority. In *L'animal que donc je suis* (*The Animal That Therefore I Am*), Jacques Derrida opens by acknowledging that he lives intimately with a cat, but his pet disappears pretty quickly. This swift abandonment reflects a prejudice of the Western philosophical tradition. From Descartes's lions to Wittgenstein's beetle in a box, animals illuminate the limits of human knowledge, a problem only augmented when images replace direct examination of these furtive, silent subjects.[11] As Stein insists, there are no animals here, only human ideas about them. Even as animal studies has emerged as a new field of inquiry, ambiguities raised by this subject reside in cultural and historical conditions that, in everyday practice, are anything but clear.

The essays in *Gorgeous Beasts* face this challenge directly, taking up the urgent task of finding animals hiding in a politicized forest of images. Where there are images, there are animals. From cave paintings to computer graphics, from personal intimacies to political relations, from literature to philosophy, images of animals suffuse human cultures. Throughout history animals circulate as different forms of cultural currency. These essays both recover the histories that give rise to these representations and track the desires that conjure them, awakening us to worlds unrealized, where history and desire meet to create knowledge. Seeing animals historically reveals what we know about them without really knowing it. We may love animals, live with them, handle, harness, or hunt them, but often without much awareness of how they come to our call or to inhabit our hearts. *Gorgeous Beasts* tracks the signs that lead to a fuller awareness of such things, enriching our everyday appreciation of the animals in our midst.

But signs take us only so far. Images lead to bodies, the bright and fecund bodies of beasts. Pursuing signs of animal life, these essays recover the physical register of representation. Images make animal bodies visible, but those bodies make images matter. The material virility of animal life provides human cultures with untold resources. Animal bodies come to matter in myriad ways: as food, fetish, labor power, companion, art object, science specimen, even military weapon. *Gorgeous Beasts* examines their creative efficacy in human cultures, their various uses and abuses. Several essays, for instance, ponder the fate of animal bodies as food for humans. Whether hunted, harvested, or slaughtered, those bodies sustain life through relationships that far exceed representation. They raise an ethical dilemma linked to the self-appointed human place at the top of current culinary hierarchies. Just as interpretive assumptions and biases based on class, race, gender, and sexuality have received productive scrutiny, so too is anthropocentric privilege ripe for critical interrogation. Such inquiry involves unraveling a network of values in which animals persistently figure as embodiments of human desires in the name of "natural" priorities. Giorgio Agamben, Luc Ferry, and other recent cultural theorists take animal bodies as their territory. *Gorgeous Beasts* follows suit, advancing an ethical inquiry into the cultural destinies of animal bodies.

If animals matter ethically, why not artistically? Animals increasingly occasion creative activity that was once considered uniquely human. It isn't simply that animals serve artists as objects. The golden calves and serial sharks of contemporary artist Damien Hirst reveal how drenched in lucre and violence animal art can be. Perhaps more interesting—because more *moving* (in all senses)—are artworks that arise as a direct effect of living (with) animal bodies. The work of artists Olly and Suzi is exemplary here: sharks co-create mixed media sculpture, or snakes collaborate as body painters. In these works, animals occasion artistic activity, which occurs in concert with humans. Something similar happens in the austere, thoughtful work of Mark Dion, featured in this volume. Animal bodies take many forms there: as pictures, taxidermy, toys, even vehicles (fig. 6). Dion forces a reflection not simply on animal bodies but, more forcefully, on their appropriation by human institutions and their beautiful irreducibility to such uses. Representative in this regard is his mesmerizing magpie under glass, taxidermied and perched on a heap of jewelry, presumably of its own gathering (fig. 7). Science stuffs this beast for display, true, but the beauty of its flashy feathers and the profusion of jewels beneath its feet wildly surpass its specimen status. The life of this animal body exceeds the terms of its representation. Such is the creativity of gorgeous beasts. They represent *and* disrupt human desires that would surmount them. Art such as Dion's creates companionships among animals of wildly different varieties and potentials. Such is the exhilarating promise and prospect of living—and creating—with gorgeous beasts.

Are there lessons here? Do animal bodies enhance life? The following essays answer emphatically: yes. They offer few conclusions, perhaps, but promote a variety of perspectives from which to view animals, humans, and the many vital relations among them. If there is a general aspiration at play here, it is that living (with) animals might become a condition of cohabitation for us today. As cohabitants, animals conjure images. They incite knowledge. They configure relationships that yield companionship. They inspire ethics. They create art. "Becoming animal," in the influential provocation of Gilles Deleuze and Felix Guattari, puts these cohabitations into motion and lives with the results.[12] *Gorgeous Beasts* thus examines animals as kith and kin, with the cunning of serpents and the innocence of doves.

Even, at times, *as* foxes.

Bestiary

The essays gathered in this volume share a common concern with the cultural fecundity of animal bodies, but they speak from a wide variety of critical perspectives and historical contexts. The scientific study of animals is a surprisingly complex practice. In chapter 1, "Animal Subjects: Between Nature and Invention in Buffon's Natural History Illustrations," Joan Landes looks at the role that images play in the emergence of animal science. She begins in a surprising place: with Picasso's illustrations of the comte de Buffon's monumental *Histoire naturelle.* Picasso's images are striking for their abstract simplicity. In this they resemble those that originally accompanied Buffon's eighteenth-century text, but for unexpected reasons. Buffon famously banished the mythical and the fantastic from the scientific treatment of animals. Scrupulously describing the austerity of his scientific method, Landes divides Buffon's devotion into three principles: exact description of animals, comparison among their varieties, and

6 Mark Dion, *Mobile Gull Appreciation Unit,* 2008 Folkestone Triennial. Photograph © Thierry Bal.

their utilitarian proximity to humans. The upshot of Buffon's naturalism was the depiction of animals less in their individuality than in their typicality, a sensibility that appealed to Picasso, who also sought to render animals in the abstract terms of an idealized type. Buffon's monkey and Picasso's bull share the typicality of their kind. Most profoundly, then, Landes's essay advances a brief history of the naturalization of animals, their relocation from the realm of fantasy to that of natural science. In a neat finishing twist, she discovers the return of the repressed fantastical bodies in the beasties and cherubs at play in images framing the animal engravings accompanying Buffon's *Histoire.*

In "Renaissance Animal Things," the second chapter in this volume, Erica Fudge undertakes a courageous reversal of commonplace notions of Renaissance humanism. Working from the assumption that animals exist neither as objects nor as representations but as *things* capable of agency, she examines their effect on Renaissance life, in particular the intimate dramas of love. As active presences in the world, co-agents of the everyday, animals participate in life completely. But they do so in a peculiar manner, which Fudge captures in the phrase "animal-made-object," signifying both an object made from an animal and an

7 Mark Dion, *Portrait of a Collector,* 2004. Taxidermic magpie, glass bell jar, wooden pedestal, miscellaneous objects, 148 x 30 cm. Private collection.

animal viewed as an object. Animals remain inseparable from the products they make possible, and this typically unacknowledged fact gives them a presence in the very products that seem to outlive them. Thus gloves made of dog skin, such as those of which Antonio Pérez writes to Lady Penelope Rich in the late sixteenth century, are not just signs of devotion but *skins* whose materiality calls doggedness to the cause of love. Similarly, the scent of the civet is more than a smell. It implicates its human wearer in animal life. Finally, at its most fearsome, the animal comes to denude the human, as in Shakespeare's *King Lear.* The animal-made-thing undoes Lear's privileged humanity, equating the breath of his beloved Cordelia with that of "a dog, a horse, a rat." As living *things,* animals are cohabitants of a life that exceeds human reckoning.

These assumptions are directly challenged by the startling use of dogs as weapons of terror, as discussed in chapter 3, "The Cujo Effect." Here, Paul Youngquist explores the transgressive transformation of a pet animal into an instrument of unspeakable horror. In the 1790s, the British decided to quell the Maroon uprising in Jamaica by deploying bloodhounds, trained by the Spanish to attack and eat black men, women, and children. Unleashed as instruments of war, the dogs were so repugnant that one British officer, witnessing such an attack on the neighboring island of St. Domingo (now Haiti), censored his own account, believing that to narrate the event accurately would be obscene. Translating the event into words would be to condone an act fundamentally incompatible with civilization, exposing the belief that white men were morally superior to black "savages" as a myth. But even the image of hounds tearing into infant flesh gives the lie to that delusion, for those dogs are extensions of human hands. Their sadistic acts reflect the perverted will of their masters. In *The Animals Issue,* philosopher Peter Carruthers makes the controversial case that animals have no direct moral claim on humans. Instead, he argues, human mistreatment of animals results in moral injury to the perpetrators, and that is reason enough for humans to treat all creatures well. The objections raised by the same British officer to using bloodhounds to terrorize Maroons in 1795 reverberate strangely with the deployment of military dogs today to terrify and humiliate enemy combatants, most infamously at Abu Ghraib prison in 2003 and 2004.

In chapter 4, "On Vulnerability: Studies from Life That Ought Not to Be Copied," Ron Broglio explores other venues in which the animal is consumed and classified through violence. The setting here is late eighteenth- and early nineteenth-century Britain, where animal slaughter and harsh working conditions were a feature of the rural landscape banished by those who wished to present a bucolic vision of rustic labor. Juxtaposing peasant poetry, visual and literary picturesque accounts of the English countryside, and agricultural treatises, Broglio ponders how the apparatus of death was "tastefully" omitted from public view. "Not to be copied"—a phrase that appears in an 1806 encyclopedia of images of the daily tasks of rural peasant life—is an injunction against viewing the act of animal slaughter. Without simplifying the connections between human and animal suffering, Broglio explores the link between labor, livestock, and English liberty, and posits the shared vulnerability of tenant farmers and animals under conditions of bare subsistence. In the process, he reveals a great deal about the social class order underpinning British wealth, starkly separating the lives of the rural gentry and urban elites from the poor.

Social class divisions were further compounded by emerging racial hierarchies during this period, as shown by Pierre Serna in chapter 5, "The Rights of Man and the Rights of Animality at the End of the Eighteenth Century." In revolutionary France even more dramatically than in England, owing to the Declaration of the Rights of Man and of the Citizen of 1789, new questions arose as to how to expand or limit the universality of rights, not only for all members of the human race but implicitly—as some began to discern—for all sentient beings. Are animals political subjects, Serna asks, and, if so, what would such a claim imply? "Animality" raises the broader question of our own animal nature, which Serna traces in philosophical and scientific queries about the nature of life. Opportunities for the expansion of rights contracted between 1795 and 1799 during the more conservative years of republican rule under the Directory, and especially following the imposition of military rule by Napoleon in 1799. In the French colonies, Napoleon reimposed slavery after it had been abolished in 1794 during the radical phase of the French Revolution. At home, democratic rights were curtailed and class divisions hardened. In the increasingly prestigious scientific fields of natural history and comparative anatomy, Serna charts a parallel movement from more egalitarian to more hierarchical views. Educational reformers under the republic and the empire aimed to place knowledge at the service of the nation, and science was accorded a privileged place within republican and imperial institutions. Ultimately, however, hierarchical principles of harmony, order, and utility in the life sciences foreclosed possibilities for more egalitarian thinking about humans, animals, and nature as an ensemble of similarities and differences. Paradoxically, too, the most advanced democratic thinkers sometimes sacrificed the rights of animals in their effort to expand the liberty of men, irrespective of their race or class.

What of animals in cages? For all intents and purposes, zoo animals are tame, radically altered by their unnatural setting. Yet they are still coded as "wild," for the state of being wild has become, as Harriet Ritvo notes in chapter 6, "Calling the Wild," more "a matter of assertion than of description." Zoos provide institutional proof of the currency of such assertions, for the idea of wildness is remarkably knotty, tied up with law, literature, and aesthetics. It is also linked to the history of scientific classification, which now ties the assertion of an authentic "wild" ancestor to the genetic preservation of animal kinds. Such assertions are inevitably political, and for animals, their consequences can mean death. Starting with the gray wolf and ending with the domestic dog, Ritvo examines the epistemological conditions of the fraught history of human encounters with animals. Her essay surveys "wildness" as connected to cultural myths of the environment, while illuminating the difficulties of understanding the condition of being wild. Is it a state, a species, a location, or a behavior? The problem she parses is lodged in language, the tool scholars and scientists necessarily use to make sense of their subject. The very act of writing about animals contains them as surely and insidiously as the bars of a cage. Is a zebra a white animal with black stripes, or a black animal with white stripes? The question becomes ironic when posed as black words on a white page.

Nigel Rothfels analyzes trophy hunting in chapter 7, "Trophies and Taxidermy," suggesting that the theatrical staging of human events, or the apparatus of the play, provides a cultural corrective to a fearful state of uncertainty. His chapter might be subtitled "101 Uses for a Dead Elephant." Insofar as "a dead elephant is not, in fact, the sort of thing one can do anything with terribly easily," Rothfels's approach offers a welcome note of absurdity. He makes the point that elephants have long been sought by trophy hunters, typically men of power and privilege who pursued them, shot them, killed them, and stuffed them, all in order to be photographed with them, thereby constructing a superior self-image while selling a personal mythology to the masses. Hunted, the elephant is "wild" merely because the narrative demands it. Rothfels opens his chapter, however, by discussing an installation video, *Play Dead; Real Time,* by artist Douglas Gordon that features a trained elephant "playing dead" and performing other tricks. Inside the gallery setting, the elephant is not wild. Rather, it embodies its own secret memories, even as it serves as a cultural repository. Not comic so much as "funny," Gordon's video radically separates the animal from its wild origins and makes it perform in a way that is subversive or "queer," knowing full well that the sexualized mythology of game animals enhances their mystique. Even their deaths become an occasion for reflection. Rothfels closes his essay with a stirring meditation on trophy photography as a medium of doubt in which a dead or dying animal provokes the human attempt to make sense of mortal violence.

The vexed legacy of scientific encounters with animal life is far from resolved, as Sajay Samuel and Dean Bavington demonstrate in chapter 8, "Fishing for Biomass." Addressing the dramatic collapse in 1992 of what had been the world's largest ground fishery in the waters off the coast of Newfoundland and Labrador, Samuel and Bavington link the disappearance of the wild codfish to the elimination of a nearly five-hundred-year-old way of life among the fishing communities of eastern Canada. With chilling lessons to be drawn for global practices of overfishing the oceans and destroying wetlands around coastal communities, the authors describe how the same techno-scientific management of wild codfish, fishermen, and the oceans, tied to capitalist exploitation of the fishing grounds, is paradoxically still being offered as the primary solution to the problem it created. Scientific representations of fish as populations or biomass, of fishermen as labor or rational economic actors, and of oceans as private or public property were harnessed to increasingly potent fishing technologies such as gill nets and bottom draggers. After more than a century in the techno-scientific crucible into which they disappeared, the once wild codfish has been resurrected as industrial biomass. Simultaneously, fishermen have been transformed into professionalized harvesters and the seas transmogrified into watery farmlands. In contrast to what they term the "algebraic machines of capital and techno-science," the authors expose the limits of knowledge systems tied to classification and marketing of fish in the commodity form of biomass, which denigrates the vernacular ways of fishermen, who once respected the fish.

Chapter 9, "Daniel Spoerri's *Carnival of Animals*," Cecilia Novero's probing exploration of neo-avant-garde artist Daniel Spoerri's animal assemblages, *Le Carnaval des Animaux* (1995), brings us full circle: back to the early modern period by way of Spoerri's recycling of the influential artist and art theorist Charles Le Brun's comparative physiognomic drawings of animal and human heads, Swedish naturalist Carl Linnaeus's classificatory systems of animals, and Darwin's notion of evolutionary time. Spoerri's *Carnival* animals, as Novero reveals, take flight through the undoing of distinctions that separate humans from nature and from animals. Spoerri's work calls upon the centuries-old popular resources of Carnival, where laughter and black humor are tied to strategies of inversion (of high and low, human and animal, death and rebirth). By decontextualizing quotidian articles like coat racks, fur coats, mementoes, folkloric masks, scientific instruments, and taxidermic animal bodies, Spoerri challenges the protocols of representation, perspective, and illusion that secure the privileged domain of high art. In the carnivalesque moment where time and space are interrupted, the possibility is realized for undoing the images of animals that humans have created to surmount their own limitations and fears.

The final section of *Gorgeous Beasts* presents the reflections of artist Mark Dion on animals and the desires and histories that shape human encounters with them. As a kind of coda to the volume, this section puts into play many of the issues raised here—with the studious playfulness that characterizes so much of Dion's oeuvre. Whether the head of a rhinoceros, a plastic-wrapped bear in a packing crate, tarred animal bodies hanging from a dead tree, or a toy bear holding a boom box in its mouth, Dion's images of animals always give way to complex networks of the desires and discourses that make them visible. Animal bodies appear as nodes of value in cultural systems they both vindicate and indict: relics in transit, specimens on display, victims of disaster, even forms of entertainment. The "marvelous museum" of Dion's work wrests unexpected meanings from animal bodies.[13] They come wrapped in history and desire, and the joy of almost any Dion installation or exhibition is the opportunity it provides to reflect on their cultural vitality. In the interview that accompanies his images,

Dion addresses many of the same concerns that the authors of *Gorgeous Beasts* explore, but from the perspective of a contemporary artist. Historical curiosity about the wildness of the wolf, for instance, burgeons into contemporary critique of its commodification. The past falls artfully into a present wholly configured by capital and techno-science. Dion's art and reflection, however, provide grounds for hope that human encounters with animal bodies can change lives for the better and revitalize the world.

Such is the hope of the essays gathered in *Gorgeous Beasts.* They find promise in animal encounters like a recent one with a whale named Tilikum. On 24 February 2010, news outlets reported the shocking death of an experienced forty-year-old female trainer caused by a twelve-thousand-pound orca, or so-called killer whale, at the SeaWorld park in Orlando, Florida. Tilikum, a male or bull whale, had lived at the park since 1992, one of eight orcas at that location. The incident marked the third time the animal had been implicated in a human death, and on this occasion numerous audience members witnessed the event. A SeaWorld spokesman acknowledged Tilikum's involvement in the 1991 death of a trainer after she fell in the pool at Sealand of the Pacific, near Victoria, British Columbia. In 1999, SeaWorld security found the body of a man—who either jumped, fell, or was pulled into frigid water and died of hypothermia—draped over the whale. In the latest incident, the female trainer also drowned, after being thrashed in the water by Tilikum (nothing "carnivorous" occurred). There were conflicting initial reports of the event. What is more, one spectator "said he heard that during an earlier show the whale was not responding to directions. Others who attended the earlier show said the whale was behaving like an ornery child."[14] Yet it was later confirmed that although the trainer was pulled underwater by her long ponytail, she was "interacting" with the whale in the water when the incident occurred.[15]

Alongside the terrible loss suffered by the friends, family, and colleagues of the dead trainer, this event points to larger cultural narratives about wild animals used in circuslike routines and their exploitation for profit. It poses the question of animal nature, of what it means to train, breed, and display a wild animal. It asks whether an undomesticated animal's behavior can legitimately be described as infantile disobedience. It raises the issue of where animals are—and ought—to be found: in this case, in a containment cage or a vast ocean, performing according to a demanding schedule, or swimming freely? And it invites speculation on the implicit violence accompanying the ownership and control of animal life. Even the name "killer whale" calls up the violence against which humans measure their superiority to animals.

The Florida park where the incident occurred was owned by the Blackstone Group, a private equity company that also owned part of the Universal Orlando theme park.[16] As the accounts further attest, the whale was purchased from a Canadian facility. As such, Tilikum was a commodity bought and sold within a global trade network that furnishes "wild" animals from marine and land habitats to parks, zoos, aquariums, and even private owners. Certainly, some members of the public uttered facetious disappointment at having missed witnessing firsthand the raw spectacle of the trainer's death, while others cleverly called this animal a "serial killer whale," or more gruesomely regretted not having "fileted and fast food[ed] it" upon its first kill, proposing "Fish sticks for everyone!" or "Sushi time?" However, most Web posts defended the whale and strenuously condemned those who would force a "wild" animal to submit to a circus routine. Concern was also expressed about the inability of trainers to interpret the animal's truculent behavior as a sign of resistance to exploitive show routines, or even as an expression of psychological malaise or physical illness.[17] Paradoxically, the orca was defended for acting both according to its nature (presumably "killing") and against its nature (performing routines). As one blogger put it, "I don't know of any cases of humans being attacked by killer whales in the ocean." Indeed, the majority of the commentators seemed to echo one writer's poignant call to "please put the whale back into the ocean where it belongs!!! FREE TILLY," or another's conclusion that "it's not the whale's fault, it's ours, Humans." Doubtless for their own reasons, not least their substantial investment in the animal, SeaWorld officials subsequently announced that they would spare the life of Tilikum. The chief of animal training at SeaWorld parks stated that "Tilikum would not survive in the wild because he has been captive for so long, and that destroying the animal is not an option either, because he is an important part of the breeding program at SeaWorld and a companion to the seven other killer whales there."[18]

The whale's saga did not end there, however. Following a yearlong hiatus after what is described as "his last killing," in March 2011 Tilikum was returned to performing at SeaWorld Orlando, too valuable a commodity to be allowed to retire. That Tilikum and his trainer shared the enclosed space of a containment tank was not an accident. It was a product not only of the intersecting histories of this animal and this particular woman but also of the intersecting history of this species of cetacean (the orca is the largest member of the dolphin family) and the human species—all of which created the conditions for an animal show before a paying public, eager to see with their own eyes a beast that has gone by other names in our cultural imaginations and belief, from the biblical sea monster Leviathan, to the ruler or custodian of the sea and benefactor of humans, according to indigenous peoples of the Pacific Northwest.

Whether viewed as those of a sea beast or of an otherworldly benefactor, Tilikum's actions appear as a failed attempt at cross-species communication and kindness, of which the sixteenth-century French author Michel de Montaigne thought animals capable. As he famously remarked in "An Apologie of *Raymond Sebond,*" "By one kinde of barking of a dogge, the horse knoweth he is angrie; by another voice of his, he is nothing dismaid. Even in beasts that have no voice at all, by the reciprocall kindnesse which we see in them, we easily inferre there is some other meane of entercommunication: their jestures treat, and their motions discourse. . . . Silence also hath a way, Words and prayers to convay."

Animal Subjects

Between Nature and Invention in Buffon's Natural History Illustrations

JOAN B. LANDES

Chapter ONE

At the invitation of the French art dealer Ambroise Vollard, Pablo Picasso undertook in 1936–37 the creation of a suite of thirty-two sugar-lift aquatint prints to illustrate a proposed modern edition of excerpts from the eighteenth-century monumental work by the comte de Buffon, *Histoire naturelle,* originally a thirty-six-volume set of scientific writings and engravings on animal and human life in nature.[1] Following Vollard's death in 1939 in a car accident, Martin Fabiani, the dealer's friend and protégé, acquired the rights to the project and published the book in the occupied Paris of 1942. In 1957 another edition appeared, based on a copy the artist originally gave to his lover, Dora Maar, which Picasso decorated with forty original drawings.[2] On the frontispiece of this edition is Picasso's playful inscription in words and pictures, where he draws Maar in the body of a bird—recalling the association of the Sirens of Greek myth (originally three dangerous bird women) with seduction and death (fig. 8).[3] Underscoring the menacing aspects of Dora's playful Sirenlike image, Picasso follows his dedicatory drawing with a page of Medusa heads, including a Medusan skull,[4] as well as another series of frightening winged skeletons, which are nothing less than skeletons of sphinxes (mythical bodies in anatomical poses), accompanied by a cracked fossilized egg and a human skull—all drawn over the page assigned in Buffon's original to the horse (*le cheval*).

8 Pablo Picasso, *Dora Maar as Sphinx.* Frontispiece of Pablo Picasso and Georges-Louis Leclerc de Buffon, *40 dessins en marge du Buffon* (Paris: Jonquières, 1957). © 2012 Estate of Pablo Picasso/Artists Rights Society (ARS), New York.

The Buffon prints are considered to be among Picasso's most important graphic productions. These are prized examples of *livre d'artiste* or *livre de peintre*—typically, "luxuriously produced, limited-edition book(s) illustrated with original prints, which are conceived as integral creations rather than as literal illustrations or simply decorative embellishments subordinate to the text." Picasso was especially enamored of the genre, creating artwork for 156 such publications.[5] Even so, this particular example of Picasso's oeuvre stands out for another reason: the manner in which it bridges invention and naturalism. As two critics remark, "Just as Picasso never made a bestiary [a medieval moralizing encyclopedia of animals], so he never really illustrated a natural history—at least not according to the rules of scientific interest or perfect naturalism which we might expect of a Stubbs or an Audubon." Yet, referring to what they call Picasso's "bastardised version" of a natural history, these same commentators underscore the naturalism of Picasso's Buffon illustrations: "The relative naturalism of the Buffon prints is a rare exception in Picasso's oeuvre. There are some early drawings showing a genuine interest in accuracy, but on the whole, unlike Leonardo da Vinci, Dürer or Stubbs, Picasso stuck to the aim of invention rather than documentation."[6]

However, while the text in a typical artist book is generally believed to "trigger" the creative collaboration between artist and publisher, it remains uncertain whether—or at least when—Picasso might have seen an example of the original work. Nor can we say whether he was acquainted with illustrations for earlier editions of Buffon's masterly work. Moreover, it is reported that there are considerable discrepancies between the existing prints and the only abridged text of Buffon that could be found in occupied Paris by the wife of the lithographer Roger Lacourière, in whose studio Picasso produced these works.[7]

In short, Picasso's deceptively simple work offers a fruitful starting point for this investigation into the work of illustration in eighteenth-century natural history texts. Whether "bestiary" or "bastard natural history," his Buffon drawings straddle the opposing categories of naturalism and imagination, description and invention, all of which have challenged artists engaged in the activities of observing, anatomizing, or representing a natural object. At times, too, Picasso pushed the parameters of the object beyond scientific inquiry to decorative amusement.[8] On a more somber note, however, Picasso's Buffon project coincides with the political upheavals in Spain, Franco's military mutiny in July 1936, and the full-scale onset of war throughout the continent and in the Far East. Indeed, his "bestiary" is not unrelated to the corpus of work he undertook during this period of intensifying gloom—most famously in *Guernica,* painted for display at the Spanish Pavilion at the 1937 Paris World's Fair. One album of drawings opens with grim sketches of a flayed sheep's head. In their starkness, Steven Nash writes, "these works relate back to Goya's still-life paintings of butchered animal parts and Cézanne's compositions of skulls";[9] and, I would add, they also reference the cold anatomical stare of natural history investigation. In fact, there is a long-standing association of the butcher, the anatomist, and the executioner, but that is another story.

The marginalia in Dora's copy of the Buffon illustrations underscore Picasso's grim mood during this period. Certainly, hybridity and monstrosity appear throughout his oeuvre. Here, as in the series of sphinxes with piercing eyes and talons, they are given an especially menacing cast. Faced with the full horror of man's actions against man, Picasso returned the animal (and human) body to the fantastic realm of myth that Buffon and other naturalists of his age had sought to expunge from the proper study of nature.

Beyond Myth: Describing Nature's Body

By the time Buffon undertook his study of nature, embracing animals, plants, and minerals, as the *Journal des Sçavans* announced on the eve of the first volumes' publication in 1749, naturalists had already done a great deal of work to clear away the mythical and fantastical presentations of animals, deriving particularly from the custom of combining real and imaginary animals in medieval bestiaries and illuminated scriptural manuscripts.[10] In pre-Renaissance works, as S. Peter Dance observes, "well observed birds, quadrupeds, insects and other creatures mingle unconcernedly with dragons, griffins, harpies, hydras, unicorns and other monstrous figments of the imagination as though they were all equally real or equally fictitious."[11]

Medieval traditions were slow to die. Another prospectus, in the October 1748 issue of *Mercure de France,* claimed that Buffon would cover "mythical" animals, among other things. "While it seems unlikely that he ever planned to do so," in the estimation of Buffon scholar Jeff Loveland, "the prospectus sheds light on the fundamentally literary expectations many readers had for natural history."[12] To be sure, the attractiveness of fantastical bodies coexisted with the rise of the new science, as is evident in the Swiss naturalist and humanist scholar Conrad Gessner's *Historiae animalium,* considered the founding text of modern zoology, and certainly one of the early modern age's most popular natural history texts.[13] Gessner's approach was typical of many other traditional natural histories, its goal to amass "encyclopedic digests of written sources on animals." In this case, as Peter Harrison points out, the text drew upon "more than 250 Greek and Latin authors."[14] Nor was Gessner alone in mixing description with ancient and later medieval accounts, or in his fascination with monstrous forms.[15] As late as the eighteenth century, natural histories assumed the existence of fantastic beings. Thus, in his *Oud en Nieuw Oost-Indiën* (1724–26), François Valentijn glamorized an illustration from Louis Renard's *Poissons, ecrevisses et crabes* (1718–19), depicting a mermaid alongside real marine animals.[16]

Buffon was well aware of the deficits associated with the promiscuous use of mythical subjects. He offered faint praise to another of the greatest naturalists of the early modern age, the Bolognese Ulisse Aldrovandi (1522–1605), for his hard labor over sixty years and for his deep knowledge of the subject. Yet Buffon advised that Aldrovandi's immense volumes on natural history could be "reduced to one tenth their present size if all those things which are useless and foreign to the subject were removed. . . . Historically, he is less adequate, often mixing in the fabulous and giving evidence of quite a penchant for credulity."[17]

In contrast to works in which fable and scientific description mingle, a stronger precedent for Buffon's approach is provided by the architect, physician, and founding member and chief anatomist of the French Academy of Science Claude Perrault in his *Mémoires pour servir à l'histoire naturelle des animaux* (*Memoirs for a Natural History of Animals*).[18] Rather than highlight the prodigious or the strange, Perrault looked for consistency and order in nature. For this reason, his illustrations have been seen as embodying "the Cartesian metaphor of stripping objects and seeing them naked."[19]

But Perrault's approach cannot be reduced merely to the axioms of Cartesian reason. Beginning with Vesalius, anatomists (and vivisectionists) led the way in the empirical investigation of the animal body, which also often involved rather gruesome experimentation on live animals. Indeed, experimental method and patient observation of the visible and invisible world—whether by the anatomist's uncovering of what the skin concealed or the microscope's revelation of objects previously unseen by the naked eye—fueled mounting objections to works of natural history, which were a mere compilation of classical authorities.[20] As a capable draftsman, Claude Perrault worked closely with the artist Sebastien Leclerc to produce drawings of both the live animals he dissected and dead specimens, with an emphasis on the accuracy of the representation. In the plates accompanying the articles on the lion or the Old World lizard (the chameleon), for example, we see the live animal intact alongside its dissected parts, whose structure and function are explicated in the accompanying text.

In this context, another of Buffon's criticisms of Aldrovandi—an attitude that he claimed also prevailed among German naturalists of the age—is worth quoting at length: "I refer," he wrote, "to the vast amount of useless erudition with which they purposively stuff their works, such that . . . they appear to have forgotten what it was they had to say to you, telling you only what others have said on the matter."

> Having once conceived the plan of outlining the whole of natural history, [Aldrovandi] sits in his library and reads one after the other the Ancients, the Moderns, the philosophers, the theologians, the jurists, the historians, the explorers, and the poets; and he reads them without any other end than that of seizing upon all the words and phrases which are directly or distantly related to his object. He himself copies and has others copy down all those remarks, arranges them alphabetically, and after having filled several portfolios with notes of all kinds—often taken without scrutiny or discretion—he begins to work on a particular subject, wishing to let nothing that he has gathered go unused. Thus, when writing a natural history of cocks or oxen, he tells you everything that has ever been said about cocks or oxen, everything the Ancients have thought about them, everything that has been imagined about their qualities, their character, their courage; all the things they have been used for; all the old wives' tales about them; all the miracles attributed to them in various religions; all the superstitious stories they have occasioned; all the comparisons poets have drawn from them; all the attributes which certain people have ascribed to them; all the representations of them found in hieroglyphics or on coats-of-arms—in a word, all the stories and all the fables which have ever been noticed about cocks or oxen. How much natural history can one expect to find in this hodgepodge of writing? Indeed, if the author had not put natural history into some articles separated from others, the natural history would not be discoverable, or at least not worth the pain of searching for.[21]

Despite his protests, Buffon nevertheless found inspiration in the considerable achievements of ancient authors, writing, "Aristotle, Theophrastus, and Pliny, who were the first naturalists, are also the greatest in certain respects." Although Pliny the Elder inhabited a world populated by intermediary species between man and animals—for example, men with dogs' heads or tails[22]—Buffon praised him for transmitting "to his readers a certain freedom of spirit, a boldness of thought, which is the germ of philosophy."[23] Yet he much preferred modern science to the Scholastic appeal to ancient sources. As a great admirer of John Locke, Buffon was not inclined to guarantee the status of classical learning but rather to follow a method rooted in experiential knowledge. It was through observation that nature's secrets would be revealed; and life was to be comprehended through an understanding of physical and material processes of nature.

"Sensualist" or "sensationist" philosophy, as it was understood and widely endorsed by eighteenth-century thinkers, emphasized the sensory origin of ideas, as well as the related (and for many supporters of this concept, the infinitely more controversial) implication that the mind (or soul) was wholly derivative of material and physical processes.[24] Still, Buffon's biographer Jacques Roger insists that the scientist "was not a man to expect everything to come from a passive mind, nor did he believe that in order to see, it was only necessary to look. It was necessary to be talented, to have 'that first spark of genius, that seed of judgment' that leads to the study of nature." He demanded more than "exact descriptions and the ascertaining in particular facts."[25] In Buffon's words:

> This is, in truth . . . the essential end which ought to be proposed at the outset. But we must try to raise ourselves to something greater and still more worthy of our efforts, namely: the combination of observations, the generalization of facts, linking them together by the power of analogies, and the effort to arrive at a high degree of knowledge. . . . A vast memory, assiduity, and attention suffice to arrive at the first end [that is, the description of particular facts]. But more is needed here. General views, a steady eye, and a process of reasoning informed more by reflection than by study are what is called for. Finally that quality of spirit is needed which makes us capable of grasping distant relationships, bringing them together, and making out of them a body of reasoned ideas after having precisely determined their nearness to truth and weighted their probabilities.[26]

As for identification and description of specimens, Buffon and other naturalists of the early modern age benefited greatly from the conquest of, trade with, and scientific exploration of the New World, the opening of Pacific routes, and greater European penetration of previously inaccessible parts of the Old World like Africa, India, and China. Expanded global exchange provided European naturalists with a flood of exotic new specimens, mainly unknown to the ancients. It also increased the challenge for naturalists like Buffon and Carl Linnaeus, his Swedish counterpart and rival, to describe, compare, and classify these multiplying specimens within a universal science of nature. In this respect, too, Buffon actually saw the ancients as his predecessors: "Alexander gave orders and made quite considerable expenditures in order to gather specimens of animals, caused them to be brought from all lands, and placed Aristotle in a position to be able to observe them well."[27]

In this regard Buffon saw himself as closer to Aristotle than to his contemporary Linnaeus, especially methodologically, in that they both took the comparison of animals with humans as their starting point. Buffon praised Aristotle for gathering "historically all the facts and all the observations which bear on the general resemblances and the sensible characteristics."[28] Thus Buffon described his own approach as directly resembling Aristotle's, noting:

> The precise description and the accurate history of each thing is, as we have said, the sole end which ought to be proposed initially. So far as the description is concerned, one ought to show form, size, weight, colors, positions of rest and of movement, location of organs, their connections, their shape, their action, and all external functions. If there could be joined to all this an exposition of internal organs, the description would be all the more complete. . . . The history ought to follow the description, and it ought to treat only relations which the things of nature have among themselves and with us. The history of an animal ought to be not only the history of the individual, but that of the entire species. It ought to include their conception, the time of gestation, their birth, the number of young, the care shown by the parents, their sort of education, their instinct, the places where they live, their nourishment and their manner of procuring it, their customs, their instinctual cleverness, their hunting, and, finally, the services which they can render to us and all the uses which we can make of them.[29]

The quarrel between Linnaeus and Buffon is commonly characterized as one over classification. Buffon ridiculed Linnaean classification for not being a science but "at the very most a convention, an arbitrary language, a means of mutual understanding . . . [from which] no real cognizance of things can result."[30] In contrast, and much in line with how he understood Aristotle's approach, Buffon proposed that the naturalist describe an object historically, looking at the differences and similarities it had with all living things, as well as its relation to its environment. He attempted "to join structure and process," or, says Peter Reill, "what we would call the synchronic and diachronic, into a unified field of explanation."[31] Indeed, it would be wrong to see Buffon as opposed to systematization *tout court.* Buffon knew that it was not enough merely to accumulate facts, "to make exact descriptions": for him, the role of the scientist was to discover the order of nature.[32] Yet he also resisted the pull of another dominant strand of contemporary thought, Cartesianism, which sought to derive this order from the methods of calculus and geometry.[33]

As noted above, Buffon takes man as his starting point, as the animal who stands at the center of nature, as that being who is a knowing subject, as the one who makes science and for whom science exists. "The real is therefore what man discovers around him, as if spontaneously," Roger observes.[34] In his "Initial Discourse," Buffon interjects a fiction not unlike those favored by others in the age of sensualist philosophy, a man devoid of all concepts and all knowledge—whether Locke's tabula rasa or the kind of statue-man imagined by Étienne Bonnot de Condillac and Charles Bonnet. "Let us imagine a man," Buffon muses, "who indeed has forgotten everything, or who awakens to completely strange surroundings." Placed alone "in a field where animals, birds, fishes, plants, and stones appear successively to his eyes," this man first perceives nothing, confuses everything, but then begins to distinguish animate and inanimate matter, arriving naturally at "that first great division, *Animal, Vegetable,* and *Mineral,*" and proceeding further "to judge the objects of natural history by the connections which they have with his own life. Those which are the most necessary or useful to him will hold the first rank—for example, he will give preference in that order of animals to the horse, the dog, oxen, etc., and he will always know more about those which are most familiar to him." Next come those that are not so familiar, but that

> nevertheless inhabit the same places and climates as he does—such as deer, hares, all the wild animals. And only after acquiring all these details will his curiosity lead him to inquire into what the animals of strange climates may be like—those such as elephants, dromedaries, etc. The case will be the same with fishes, birds, insects, shellfish, plants, minerals, and all the other productions of nature. He will study them in proportion to their usefulness; he will consider them to the extent that they are familiar to him, and he will rank them in his mind relative to the order of his acquaintance with them, because that is indeed the order according to which he experienced them and according to which it is important to him to preserve them.[35]

Ordering the Animal Body

I have identified some of the most decisive protocols guiding the presentation of animals in Buffon's magnum opus: exact description, comparison, proximity to man.[36] The first quadrupeds discussed in the work are the familiar and the domesticated. Although early on a "wild" animal makes it into the preliminary order, pride of place, after man, is given to the horse, followed by the ass, the ox, the sheep, the goat, the hog, the hog of Siam, and finally the wild boar: Buffon writes of the two hogs and the boar, "I have joined these three animals, because they form but one species. The one is the wild animal, the other two are the same animal, only in a domestic state."[37] Only then do we encounter dogs and cats, which, while presumably close to the house, are given a lower rank than the preceding animals because they lack a comparable "utility." Following this array of animals most attached to the domestic economy, Buffon moves on to wild animals, and here, too, he seems to order them from the most familiar to the most strange, acknowledging that some of them are available thanks to France's colonial expansion. For example, he identifies what he calls the Guianese martin as having been "sent from Guiana to M. Aubry, curate of St Louis."[38] Following the wild and carnivorous beasts, he comes finally to monkeys (volume 8 of the original nine volumes devoted to animals, volume 9 being for additions), although Buffon acknowledges the species' anatomical proximity to man—with the important reservation of its lacking language and reason. Thus, the smaller "ourang-outang" is depicted with a seemingly human expression (fig. 9).

Already some of the peculiarities of this scheme are discernible, the most striking doubtless being that the more remote animals appear to be those in furthest proximity from polite society, the audience for whom Buffon wrote. Even here the issue is what we might call the animal's natural environment, for certainly society's members kept various exotic animals not native to western Europe as domestic pets. There was a virtual craze for canaries and parrots during the eighteenth century, and owners could avail themselves of a handbook first published in 1705 and reprinted many times thereafter: *Nouveau traité de canarie* by J.-C. Hervieux de Chanteloup, a timber inspector and also "governor of the canaries of Mme la princesse De Condé."[39] As early as 1597 a traveler to Marseilles "watched foreign animals being unloaded onto the crowded docks and reported seeing a pet ostrich at a country house; a chained leopard (which had supposedly killed seven people a few weeks before) in a town square; a lion cub at an inn; a porcupine in a private home; and, at the duc de Guise's mansion, a large monkey named Bertram."[40] At the end of the eighteenth century, more and more exotic animals were available to people further down the social scale; and ports like Le Havre and Marseilles were reported to be brimming with birds. The African slave trade was a boon to the trade in animals, as Buffon acknowledged in his article on the rose-ringed parakeet: "the ships that leave from Senegal for Guinea, where this parakeet is commonly found, bring quantities of them along with the Negroes to our islands in America [the Antilles]."[41] In addition, although never as widespread as birds, "monkeys also became more available in the latter part of the eighteenth century. A book full of monkey anecdotes from 1752 stated that they had recently become quite common and that one encountered them in Paris 'at the homes of a relatively large number of people.'"[42]

9 [Juste] Chevillet and Georges-Louis Leclerc, comte de Buffon, *Le Jocko,* 1766. Engraving. In Buffon, *Histoire naturelle, générale et particulière, avec la description du Cabinet du roi,* 44 vols. (Paris: Imprimerie Royale, 1749–1804), 14:81.

Picturing the Animal Body

The tremendous display of nature's variety that we find in Buffon's great works, and the spirit of admiration or marvel undergirding the representation of animal forms and structures, seems wholly at odds with the complaint lodged by William Wordsworth against eighteenth-century naturalists, that is, that they "murder to dissect" the works of nature.[43] As for the picture, it is of some interest that Linnaeus, the greater systematizer, had little regard for the image. In 1717 he wrote, "I do not recommend the use of images for the determination of genera. I absolutely reject them—although I confess that they are more pleasing to children and those who have more of a head than a brain. I admit they offer something to the illiterate. . . . But who ever derived a firm argument from a picture?"[44]

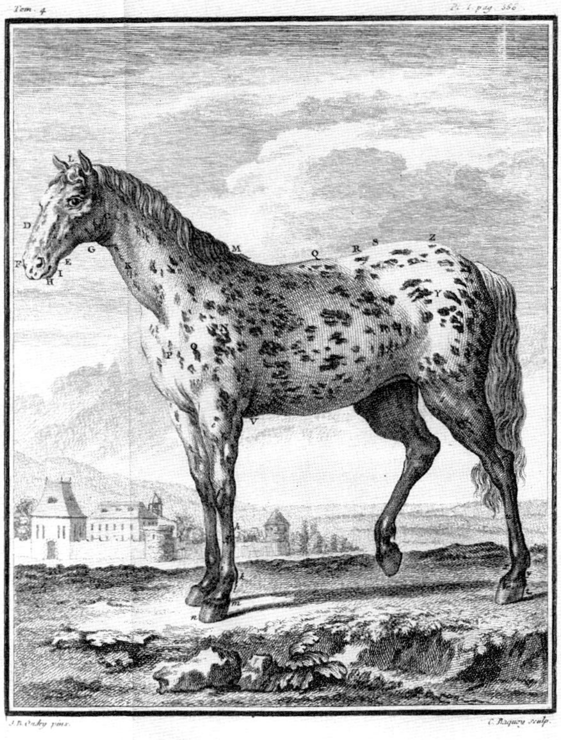

10 Jean Charles Bacquoy, Jean-Baptiste Oudry, and Georges-Louis Leclerc, comte de Buffon, untitled [The (Domesticated) Horse], 1749. Engraving. In Buffon, *Histoire naturelle,* 4:366.

Apparently, Buffon, for one. Drawing upon heterogeneous sources, including live animals in different circumstances, mounted, stuffed, and anatomized specimens, older natural history illustrations, and verbal descriptions, he and the artists under his direction undertook a very thorough visual record of every animal discussed in the text.[45] Indeed, Buffon's rejection of Linnaeus's arid scheme of classification in favor of groupings based on "resemblances and differences . . . derived not just from one part (of an animal), but from the whole ensemble" was tied directly to his methodological appeal to systematic empirical description and classification. Unlike naturalists whom Buffon accused of imposing a "barbarous" scheme, his approach would produce the very order of knowledge that it sought to understand (see, for example, fig. 10).[46] As Alex Potts explains, "comparison and description constituted part of the same process. It was in the act of description that the forms of an animal's features were clearly defined in a way that made possible systematic comparison. Equally comparison

11 Jean Charles Bacquoy, Jacques de Sève, and Georges-Louis Leclerc, comte de Buffon, untitled [The Bull], 1766. Engraving. In Buffon, *Histoire naturelle,* 14:531.

highlighted the features that exact description needed to take into account if it were to define the visible appearance of the animal fully and clearly. Description then was an essential feature, not just of the recording of knowledge, but also of its articulation."[47]

And what was being recorded was not really the individual animal as such but rather its ideal or typical form. Yet unlike the neoclassical artist who also sought to produce a "beautiful type" by synthesizing observations from several models, what is emphasized here is the animal as representative of its species.[48] Thus the viewer will be aware from his or her own experience of the variety among actually existing horses but will simultaneously be expected to understand this image as a condensation of everything typical and familiar about the species. At stake for Buffon is a "general knowledge of all animals." Or, as Potts says of the illustration of the bull (fig. 11), "the detailing is obviously intended to be naturalistic, while the clearly marked outline bespeaks the more diagrammatic aspect of the image, its conception as a partly schematic articulation of the characteristic forms of the animal."[49]

For all the insistence on an accurate drawing of the animal body, there is something incredibly stilted, rigidly schematic, about the Buffon images.[50] The artist has chosen to stage a scene, sometimes setting the animal against a classical or "foreign" background—for example, the elephant is depicted against Asian pagodalike structures, the hairy armadillo against ancient ruins, with Corinthian columns and distant palm trees—or with certain social accoutrements attesting to the privileged status of the would-be owner of such an animal, as in the images of pugs in privileged aristocratic settings. Or the animal (or animal skeleton) is posed on a plinth, as if in a natural history cabinet, or before a drawn curtain, as in a dramatic or anatomical theater.[51] If Wordsworth's charge is not entirely fair, what of the almost commonplace nineteenth-century argument that "true picturing and description" is an "unmediated seizing upon facts and raw realities," not a stilted and ultimately "artificial or unnatural system of grid"?[52] From that perspective, Buffon and his artist collaborators (including the illustrators Jacques de Sève and François Nicholas Martinet or the engraver Jean Charles Bacquoy) were no different from Linnaeus: in practice, Buffon's paradigm was "the antithesis of the naïve eye theory of accurate observation."[53]

Diagrammatic Bodies: Nature or Invention?

Despite the cultural coding of the image, is there still not something compelling about the condensation of the schema, or, to resort to hopelessly outdated language, something "true" about the body as represented in such a picturing? It is precisely this aspect of Buffon's illustrations that is captured so brilliantly, I would suggest, in Picasso's "quasi-naturalistic" natural histories. Yet Picasso's sympathetic image of the monkey also seems to undercut the anxiety that Buffon expressed about the possible similarity between monkey and human (fig. 12). In the text published alongside Picasso's print, "we are warned not to be deceived by the monkey's apparent closeness to man. In fact, for the eighteenth-century naturalist, the monkey is far from being the most intelligent animal: 'Just as its nature is lively, its temperament is hot, its nature petulant, so none of its feelings are mitigated by education. All its habits are excessive, and more closely resemble the actions of a lunatic than of a man.'"[54]

In the bold outlines of this jarringly unstereotypical monkey body, Picasso nevertheless condenses the characteristics of monkeyness. For this is not a monkey performing monkey acts or ostentatiously imitating human behavior, nor is

12 Pablo Picasso, *The Monkey (Le singe),* 1936, published 1942. Aquatint and drypoint. From the illustrated book *Eaux fortes originales pour textes du Buffon (Histoire Naturelle)* (Paris: Martin Fabiani, 1942). The Louis E. Stern Collection. The Museum of Modern Art, New York. © 2012 Estate of Pablo Picasso/Artists Rights Society (ARS), New York. Digital image © The Museum of Modern Art.

13 Pablo Picasso, Bull series, 6 states, 5 December 1945–17 January 1946. Lithograph, worked first with wash, then pen, scraper, and crayon, on stone. © 2012 Estate of Pablo Picasso/Artists Rights Society (ARS), New York. Courtesy of the Norton Simon Art Foundation.

he shown swinging from a tree, emphasizing his species' agility. The profile view of the animal—a long-standing convention deeply rooted in Enlightenment imagery—is retained in the upper torso but not in the animal's lower torso. It is as if he would sit like we ("Europeanized" human animals) do, with legs able to swing freely from a seat bench or planted firmly on the floor. Instead, his legs are in a crouch, his knees at the height of his chest, which makes him all the less human. But what about his gesture? Is he offering something to a passerby, or even to the viewer? Is he asking for food or for contact, perhaps a handshake? And what, ultimately, of the genital display? Tamed or contained, hirsute but heavy in stature (another mark, perhaps, of his domesticated condition?), his deflated sexual member seems to mark not so much his diminished virility as his isolation. The monkey seems to be asking, what exactly is monkeyness?

In a remarkable lithographic series undertaken by the artist at the very end of 1945, Picasso distills the image of the bulky and detailed bull to the simplest, most graceful outline of the animal—with which, it must be noted, the artist was most intimately associated in its many manifestations in his work—from the raging bulls of the *corrida* to the mythical Minotaurs and the dreamlike animals in the Neolithic wall paintings of southern France and northern Spain (fig. 13). The reminiscence of Fernand Mourlot, in whose lithographic workshop Picasso produced these works in a feverish period of four months, points to the marvel with which his technique and approach were viewed by those around him.

> One day . . . he started work on the famous bull. It was a superb, well-rounded bull. I thought myself that that was that. But not at all. A second state and a third, still well-rounded, followed. And so it went on. But the bull was no longer the same. It began to get smaller and to lose weight . . . Picasso was taking away rather than adding to his composition. . . . He was carving away slices of his bull at the same time. And after each change we pulled a proof. He could see that we were puzzled. He made a joke, he went on working, and then he produced another bull. And each time less and less of the bull remained. He used to look at me and laugh. "Look . . . ," he would say, "we ought to give this bit to the butcher. The housewife could say: I want that piece, or this one . . ." In the end, the bull's head was like that of an ant. . . . At the last proof there remained only a few lines. I had watched him at work, reducing, always reducing. I still remembered the first bull and I said to myself: what I don't understand is that he has ended up where really he should have started. But he, Picasso, was seeking his own bull. And to achieve his one line bull he had gone in successive stages through all the other bulls. And when you look at that line you cannot imagine how much work it involved.[55]

Rather than begin from the simplest elements and work toward the richest and most complex, Picasso moved in the opposite direction, in effect violating the aesthetic rules of draftsmanship, as well as the aesthetic and cultural accretions of history—after all, it is assumed that we evolve upward, not downward. Still, it should be said, Buffon did entertain the idea of "degeneration" as a departure from nature. He also had a prescient way of conjuring up a whole animal in just a few sentences at the beginning of an article.[56] As for Picasso, it has been noted how much he admired the naïve art of small children, popular graffiti, and the qualities of African sculpture (not its "primitive" character, a feature of Egyptian or Chaldean sculpture); that is, he appreciated that they were magic things, intercessors, mediators, fetishes, weapons. "In this link between what he called 'the spirits,' 'the unconscious,' 'emotion,' and the exorcism of the accumulated legacy of Western Tradition," observes Irving Lavin, "Picasso's enterprise is phylogenetic" as well as ontogenetic.[57]

Although Picasso was probably not aware of it, there is another odd resonance between his effort to strip away cultural accretions to get to the "ultimate abstract graphic distillation of reality"[58] and what has been called Buffon's "idea of controlled regression from the known to the unknown, from the complex, to the less complex, and from experienced and intuited human relations to the related but dissimilar relations of other organic beings. . . . [Moreover,] this argument had a proviso that humans could not become detached observers of natural processes."[59] As Buffon put it,

> The first truth that emerges from this sober examination of Nature, is a truth perhaps humbling to man. This truth is that he ought to range himself in the class of the animals, which he resembles in all of his material aspects. Even their instincts will appear to man more certain than his own reason, and their industry more admirable than his arts. Then, examining successively and by order the various objects that compose the Universe, and placing himself at the head of all created beings, he will see with astonishment that one can descend by almost imperceptible degrees from the most perfect creature to the most unorganized matter, from the most organized animal to the most inert matter.[60]

In Picasso's lithographic series, each print is a "state" with its own heightened reality; each is a simplification or reduction of what preceded it. My point is not to reduce Picasso to a latter-day Buffonian, only to point to a possible parallel here between Picasso's visual methodology and Buffon's radical effort "to close the breach between mind and matter, man and animal . . . [to make what is] contingent the prime area of inquiry."[61] The medium of lithography, to which Picasso was only first introduced on this occasion, offered the artist unparalleled possibilities. Unlike the metal plate used in etching and engraving, where erasure is extremely difficult, Picasso exploited with true genius the potential of "the lithographic stone . . . for the particular story he had in mind—the retrogressive destruction of a single work of art back to its original state; or, what amounts to the same thing, the progressive evolution of a single work to its ideal state." In Irving Lavin's acute formulation, "Picasso's joke about the butcher and the housewife reveals part of what he had in mind: to retrieve the bull's constituent parts, to recover and reduce the *disjecta membra* of his dream bull—bred of pure lines—to an elemental, disembodied, quintessential bullishness." Moreover, Lavin adds, "Another insight is suggested by one of the most striking aspects of the animal's metamorphosis—duly observed, at least in part, by the perspicacious craftsman—the progressive diminution in the relative size of the head and the genitalia, surely metaphors for rationality and brutishness. Picasso's bull was headed toward a preternatural state of illuminated absent-mindedness and incorporeality—before it had acquired the bulky accretions of sophisticated European culture."[62]

What is left after all this is the essence of the bull—the bull in its magical or mythical essence, a bull not unlike the bulls drawn by unknown Neolithic artists on the walls of the caves of southwestern Europe.

Cherubs and Beasties: A Coda

I began with some speculations about myth, and the way that beginning with the Enlightenment the mythical body is exorcised, pushed beyond the frame, outside the scientific object of modern natural history. However, I don't want to leave the impression that one would have to await the reintroduction of Arcadian motifs in nineteenth-century romantic historicism, the interrogation of natural and geometrical forms by German biologist Ernst Haeckel in his *Kunstformen der Natur* (*Art Forms in Nature*), or, better yet, the expansive primitivist explorations of twentieth-century artists like Picasso—or Paul Klee, another brilliant example of the modern artist's fascination with children's drawing and the exploration of naïve styles—before the mythic imagination once again becomes, as in non-Western and ancient cultures, an imagination that bridges the gap introduced between art and science.

By way of conclusion, and perhaps as an invitation to another episode in my journey to better understand the graphic and scientific exploration of the animal body in eighteenth-century thought, I point to a strand that to my knowledge has not been fully acknowledged, and certainly has never been discussed in any comprehensive way by other scholars of Buffon's attempt to codify the

HISTOIRE NATURELLE.

L'ÉLÉPHANT*.

L'ÉLÉPHANT eſt, ſi nous voulons ne nous pas compter, l'être le plus conſidérable de ce monde : il ſurpaſſe tous les animaux terreſtres en grandeur, & il

* *Éléphant*, en Grec, Ἐλέφας; en Latin, *Elephantus*, *Barrus*; en Italien, *Leophante*; en Eſpagnol, *Elephante*; en Allemand, *Helphant*; en Anglois, *Elephant*; en Orient, *Elſil*. *Phil* ou *Fil* eſt un mot Chaldéen qui ſignifie ivoire, & dont Munſter s'eſt ſervi pour déſigner l'éléphant. On appeloit autrefois l'éléphant *Barre* aux Indes orientales; & c'eſt vrai-ſemblablement de ce mot qu'eſt dérivé le nom *Barrus*,

Tome XI. A

14 Jacques de Sève, A.-J. Defehrt, and Georges-Louis Leclerc, comte de Buffon, untitled, 1764. Engraving. In Buffon, *Histoire naturelle,* 11:1.

enlightened description, classification, and illustration of the natural history object: that is, I look to the frame, not the picture—the rococo motif within the various tableaux that introduce the chapters of this great work of natural history (fig. 14). Here we find plump cherubs playing delightedly among the beasts of the animal kingdom, with echoes not only of classical works but also of that other great fabulistic tradition in the Christian West within which one finds the biblical story of Noah's ark. In any event, in these plates the artist Jacques de Sève allows his imagination full play in compositions that have more in common with the idyllic imagery of contemporary rococo artists than they do with the protocols of natural history illustration as it came to be understood from the eighteenth century to this day. Here the playground of nature is inhabited by all of God's creatures, from man and beast to beasties and cherubim.

Renaissance Animal Things

ERICA FUDGE

Chapter TWO

Animal Matter and So-Called Human Culture

In their *Dictionary of Traded Goods and Commodities, 1550–1820,* Nancy Cox and Karin Dannehl record "nearly 4,000 terms found used in documents relating to trade and retail in early modern Britain." The objects listed in the dictionary range from the easily recognizable "oats" and "sauce spoons" to the now lost "deadhead" (used for red pigment) and "scabious water" (good, apparently, for "diseases of the breast and lungs"). The first animal object in the dictionary sits between the opening entry, "ABC"—a child's spelling book—and the third, "absinthe," the plant wormwood.[1] It is "abortive vellum," that is, "vellum made from the skins of aborted calves or other animals. It provided a soft skin with a fine grain suitable for art work." This, in many ways, might exemplify many animal commodities: in it, the actual animal's presence seems to have disappeared, to have been overlaid by human culture.[2] But this is not always the case, of course, and some items in the dictionary offer another set of possibilities for early modern consumers in which the animal is more obviously present: the last two animal entries, for example, are "Yorkshire ham" and "young beast": dead or alive—meat or beast—these animals are commodities.

The *Dictionary of Traded Goods and Commodities* affords a fascinating glimpse into a world of stuff. This is a world that has become a key area of inquiry in Renaissance studies over the past fifteen years. Ann Rosalind Jones and Peter Stallybrass, for example, whose work has been important to my own thinking, have focused on clothes and the movement of clothing through inheritance and the secondhand market and have traced new ways of thinking about class and gender relations in Renaissance culture through this research.[3] For them, as for others, the object has displaced the text: things have supplanted language as a focus of critical attention.[4] For scholars like me who work in Renaissance studies and in animal studies, the focus on material objects rather than textual representations brings with it new possibilities but also new problems. For while, as the *Dictionary of Traded Goods and Commodities* reveals so clearly, there is an abundance of animal matter to deal with in early modern England, there is also a possibility that concentrating on animals as the stuff of the market might go against one of the central tenets of animal studies, which is that animals should be acknowledged and engaged with as active presences in the world. When this happens—when the animal is regarded as an agent—the human can be seen in a new light, as a being relying on and inseparable from animals to the extent that the human world can only be regarded as the *so-called* human world, that is, as one constructed by numerous participants, not all of which are members of our own species.[5] The danger is that reading animal matter rather than reading living beings might, in short, only reinforce the perspective that relegates animals to the realm of inert objects. Can livestock, I wonder, ever be more than *live stock*?

The first way to reconcile the apparent disjunction between an interest in animal stuff and an interest in animals as active presences in the world is to note that a living animal and animal matter are not separate categories. Like subject and object, they are utterly intertwined.[6] Thus, while it is obvious that live animals and dead animal products are not the same, assuming a distinction between, say, the cow in the field and the milk in the carton actually obscures the reality in which milk produces cows as much as cows produce milk. Denying this entanglement—maintaining the opposition of subject (active) and object (passive), animal and matter—perpetuates a system that can present us with happy cows and with commodified milk simultaneously.[7] But as well as this I want to propose—and this is the focus of this chapter—that animal matter can have an active presence in so-called human culture, too. And in response to this double meaning—the inseparability of animal and product, the potential agency of animal stuff—I propose a term that makes inseparable living animal and dead matter: "animal-made-object." This term carries two simultaneous meanings: (1) the *animal-made* object—the object constructed from an animal, and (2) the animal *made-object*—the objectified animal. Using this term might, I hope, not only remind us of the concurrent status of animals as both agents and matter but also of the nature of the relationship we have with them.

As well as my adoption of this term, I want to outline briefly two theoretical perspectives that act as important lenses through which to bring animal matter into focus in animal studies. The first, and the one most familiar to scholars in this field, is Actor Network Theory (ANT). ANT offers a conception of agency that allows us to see animals as active presences without having to assert that they have pseudohuman subjectivity.[8] Indeed, ANT proposes that humans themselves are simply actors in networks,[9] and that the concept "pseudohuman subjectivity" is meaningless because "human subjectivity" itself does not exist in any a priori sense. As John Law puts it, "The argument is that thinking, acting, writing, loving, earning—all the attributes that we normally ascribe to human beings, are generated in networks that pass through and ramify both within and beyond the body. Hence the term, actor-network—an actor is also, always, a network."[10]

In a recent essay Bruno Latour traces one way in which such ideas about the human are resisted. He notes that the relegation of technology to the realm of means rather than ends "by a large number of sociologists" is underpinned by and simultaneously reinforces a conception of who it is that we think we are. In the face of the increasing "autonomy" of technology, he argues, the response of these sociologists is clear: it is to disengage the human "from this domination by technologies" as if such separation were both simple and possible. "To become moral and human once again," Latour writes, "it seems we must always tear ourselves away from instrumentality, reaffirm the sovereignty of ends, rediscover Being; in short, we must bind back the hound of technology to its cage."[11] The canine image here is perhaps accidental—Latour says nothing directly about animals in this essay—but it is clear that his ideas about technology can be usefully applied to the role and place of that group of nonhumans. Animals, too often, are bound back, absented from the picture, made to seem unnecessary and inconsequential, with the result that the human emerges as the only necessary and consequential being in the frame. To offer a wonderfully and accidentally surreal example from the discipline of history, in his 1962 study *The Wool Trade in Tudor and Stuart England,* Peter J. Bowden refers to sheep farmers in the late seventeenth century as "wool growers."[12] The animal has been absented from this image as a sentient being with a mind of its own (let alone as an agent) and is depicted instead as a kind of invisible—and by implication virtually unnecessary—plantlike organism. And because, in this discourse, plants don't act—they simply grow—the human is represented as the only actor.

With Latour, many scholars are attempting to challenge such assumptions. Lucas D. Introna, for example, takes us toward the second theoretical perspective that I think will aid in bringing animal matter into animal studies, and offers what might have been (but wasn't) written as a direct response to Bowden's image of sheep farmers as wool growers. Introna writes:

> We most often do not consider these things that surround us beyond their instrumental value. They seem just to be there, available (or sometimes not) for us to draw upon. Lurking in the shadows of our intentional arc they sometimes emerge as relevant, become available, fulfil their function, and then slip back into the forgotten periphery of our intentional project—often doing the invisible work that was allocated them in a now forgotten time and place. In many ways we have allocated to them the role of silent workers, the décor and backdrop that constitute the possibilities of our lives, but are best forgotten. Nevertheless, as we draw on them they become more and more part of who we are, or who we are becoming.[13]

ANT serves to remind us, then, of two things: first, of the presence of all the "silent workers" that make up our so-called human world, and second, of their place in constructing who it is that we are. It serves, if you like, to bring back the abortive vellum, not as mere background but as an agent in the world.

The second theoretical perspective that I think might allow a focus on the matter made from animals to be a productive site for work in animal studies is less well known in that field, despite its often close relations to ANT. It is present in Introna's conception of the stuff that lurks in the "shadows of our intentional arc," and is thing theory.[14] According to Bill Brown, an object is perceived to possess power and agency in the world through its emergence as a "thing." "We begin to confront the thingness of objects," he writes, "when they stop working for us; when the drill breaks, when the car stalls, when the windows get filthy, when their flow within the circuits of production and distribution, consumption and exhibition, has been arrested, however momentarily. The story of objects asserting themselves as things, then, is the story of a changed relation to the human subject and thus the story of how the thing really names less an object than a particular subject-object relation."[15]

Thing theory allows us to recognize that objectified animals, that is, animals *living or dead,* can and should be read as having active presences in the world: they are "asserting themselves" (Brown's phrase) or are recalcitrant (mine)—the *OED*'s happily animalistic definition of the word "recalcitrance" is that it is a "kicking against constraint," and, in her study *Domesticated Animals from Early Times,* Juliet Clutton-Brock writes of the "erratic and recalcitrant . . . behaviour" of "solitary carnivores" (her example is the ferret) that cannot be tamed.[16] Things, in these terms, are hounds that will not be caged or domesticated. Without human intention, and indeed potentially against human intention, the animal-made-object can be seen from this perspective to construct new meanings, beings, and relationships. It is a truly active presence in the world.

My intention in this chapter, then, is to see whether Latour's idea that an orthodox—*humanist*—conception of being human requires us to "bind back the hound" of the nonhuman "to its cage" holds true in the pre-Enlightenment past as well as in the post-Enlightenment present. And it is to see whether paying attention to the recalcitrance of the animal-made-object to which thing theory alerts us might be one way in which such attempts to contain the nonhuman can be brought into view. My focus here on animal matter rather than on living animals will, I hope, allow me to argue *in extremis* for the power of the nonhuman animal to effect change upon so-called human culture even when the animal as sentient presence has been removed.

I begin with animal skins, move to human scent, and conclude with *King Lear,* a play in which furred gowns hide all, and men are forced to smell their way to Dover.

The Thingness of Skin

The recalcitrance of the animal-made-object that makes it a thing can be traced in Peter Stallybrass and Ann Rosalind Jones's study of gloves in the early modern period. In their article they trace out the many ways in which gloves are used to construct, contain, and transport political, sexual, and gendered meaning. And they turn at one point to the gloves referred to by Antonio Pérez, a Spanish exile in Elizabeth's court. Pérez refers to these gloves in an extended conceit. Writing to Lady Penelope Rich around 1595, he states, "I have been so troubled not to have at hand the dog's skin gloves your ladyship desires that, pending the time when they shall arrive, I have resolved to sacrifice myself to your service and flay a piece of my own skin from the most tender part of my body, if such an uncouth carcass as mine can have any tender skin. To this length can love and the wish to serve a lady be carried that a man should flay himself to make gloves for his lady out of his own skin." The letter concludes, "The gloves, my Lady, are made of dog's skin, though they are mine; for I hold myself a dog and beg your ladyship to keep me in your service upon the honour and love of a faithful dog." Pérez signs himself "Your Ladyship's flayed dog." Stallybrass and Jones write, "Pérez extends a trope that was already well established in classical antiquity. The lover, transformed into gloves, will always be near his beloved. When she wears him, he will be able to touch her."[17]

I do not want to challenge Stallybrass and Jones's interpretation of the glove as sexual metaphor, but I want to suggest that they miss out on a layer of meaning in Pérez's dog-skin gloves that can be found in the animal that is present in the animal-made-object. The lover wants not only to be a glove but a glove *made from the skin of a dog.*[18] As such, we might be able to see another Renaissance trope being used here: just as puns are often made between "dear" and "deer," between "heart" and "hart" (a mature male deer, or "an animal with at least ten tines")[19] in the love poetry of the period, so Pérez is presenting an animal pun here: this admirer is dogged in that he is utterly faithful to his mistress.

But in his use of the conceit of the dog-skin gloves Pérez is not only displaying his loyalty; he is also what we might term becoming doggish; he is being made canine. And this, I think, is where the gloves assert themselves against humanist intent. In short, it is where the "caging"—the utter restraining—of the hound that is present in the flaying and in the manufacture of the gloves is undone and the animal-made-object becomes a thing.[20] The dog's persistent presence in the gloves seems to work against rather than with the sexual symbolism, in that the extended conceit might be read not as a metaphor of fidelity but as bestializing—as making unnatural—the sexual innuendo and thus possibly both Pérez and Lady Rich. In this period the dog is a man's best friend: the phrase "loue me, loue my Dogge" appears in Philip Stubbes's *Anatomie of Abuses* of 1583, where Stubbes says that it is "a common saiyng [*sic*] amongest all men," for example.[21] But a dog is also a despicable, less than human creature. In *The Merchant of Venice,* which was written perhaps a year after Pérez's letter, Shylock mocks Bassanio's request for a loan, asking him:

> Shall I bend low, and in a bondman's key . . .
> Say this: "Fair sir, you spat on me on Wednesday last;
> You spurned me such a day; another time
> You called me dog; and for these courtesies
> I'll lend you thus much moneys?"[22]

I am not suggesting that the Spanish Pérez knew to represent his outsider status through the image of a dog, but I am proposing that the dog in the dog-skin gloves that he discusses can be interpreted as being recalcitrant, as kicking against the intention of the human. The putative lover uses two established tropes of love poetry—the sexualization of the glove and the lover as punning animal—but in doing so he also emphasizes the reality of the dog in the gloves, and as a result his own body becomes a carcass. As Stallybrass and Jones ask in a different context, in their essay: "Where does the skin of animal end and the skin of human begin? It is hard to tell."[23] This difficulty I read as evidence of the recalcitrance of the animal-made-object, which makes it an animal thing.

The fact that Stallybrass and Jones do not pursue their question in relation to gloves further is a pity, because animal skin (as they point out in another study) is an interesting thing in early modern thinking, because its relation to the human is a complex one. According to Genesis 3, Adam and Eve experience shame at their nakedness after the Fall and clothe themselves in fig leaves. But God replaces these: "Unto Adam also and to his wife did the Lord God make coats of skins, and clothed them" (3:21). In their book *Renaissance Clothing and the Materials of Memory,* Jones and Stallybrass recognize the ambiguity of God's "gift":

> In making clothes for Adam and Eve, God gives them his livery. His livery is both a form of protection and a threat. Fig leaves, the clothes that Adam and Eve made for themselves, are minimal and temporary. These vegetable forms will wear out. In contrast, animal skins give greater warmth and have a longer shelf life. But they also inscribe upon the bodies of Adam and Eve the first deaths in Eden. For animal skins become clothing through the deaths of the animals. In asserting their independence from God, Adam and Eve clothe themselves. In reclaiming them as his subjects, God reclothes them in his livery—a livery of protection, but also a livery haunted by death.[24]

Animal skins thus mark humans as all-powerful (animals will be killed for them) and simultaneously as all-frail (they *need* animals to be killed for them), and this paradox can be traced in two very different readings of skins in this period.

Philip Stubbes, the man who declared the commonness of the declaration "loue me, loue my Dogge," argued that God gave animal skin to Adam and Eve out of pity and that it was thus

> meane & base attyre [which] should be as a rule, or pedagogie vnto vs, to teach vs [that] we ought rather to walke meanelye, and simplye, then gorgiously, or pompously: rather seruing presente necessitye, than regarding the wanton appetits of our lasciuiouse mindes: Not withstandinge I suppose not, that his heauenlye maiesty would, that those garments of lether, should stande as a rule or patterne of necessytie vnto vs, wherafter we shold be bou[n]d to shape all our apparell for euer, or els greeuouslye to offende: but yet by this, wee may see, his

> blessed will is, that we should rather go an ace beneth our degree, than a iote aboue.[25]

For Stubbes, corruption in clothing comes in ornament and overreaching, whereas the skin of a dead animal marks on the human self-abasement and is thus the closest that fallen man can come to holiness.[26]

There is, though, another interpretation of animal skin in this period that takes as its scriptural model not God's introduction of animal death into Paradise but, I suggest, the story of divine prophecy and human failing in Genesis 25–27. In the tale of Jacob and Esau, God has foretold their mother, Rebekah, before their births that the hierarchy of primogeniture will be overturned, and that "the elder shall serve the younger" (25:23). Despite this, Rebekah encourages her younger son, Jacob, not to wait to see the fulfillment of God's will but to bring it about himself. So at his mother's entreaty he puts on Esau's clothes and covers his own smooth hands and neck with the skins of goats in order that he should be mistaken for his hairier older brother by their father, Isaac, who "was old, and his eyes were dim, so that he could not see" (27:1). Indeed, in the 1568 interlude *A newe mery and wittie Comedie or Enterlude . . . vpon the Historie of Iacob and Esau,* gloves appear once again to challenge identity. Rebekah says to her younger son:

> I haue brought sleues of kid next to thy skin to weare.
> They be made glouelike, and for eche finger a stall:
> So that thy fathers feeling soone beguile they shall.[27]

Jacob's aim is to trick his father in order to receive the blessing due to Esau, and he is thus, like Adam, another man seduced from the right path by a woman.

In this story animal skin marks out the fraudulent nature of the wearer: here Jacob is not so much a man going "an ace beneth [his] degree," as Stubbes saw the wearing of skin. He is, in his disguise, an image of human lack of faith in God. The villain of the piece, though, is perceived to be Rebekah, who doubts that God will fulfill his prophecy. Calvin wrote of her failing, at this moment: "She knewe [that] it was an immutable decree, by which Jacob was elected and adopted. Why then doth she not patiently tarrie, vntill God confirme in very deede, & do shewe that the same is ratified, which he hath pronounced from heauen? Therefore she obscuring the heauenly oracle with a lye, abolisheth so much as in her lyeth, the grace promised to her sonne."[28] Rebekah appears to have agency in that she makes a man dress in animal skin, but in exhibiting this capacity she actually undermines the very structure by which such agency exists. She questions God's authority and as such disputes the source of human power. As the prologue to the *Historie of Iacob and Esau* puts it, "For it is not . . . in mans renuing or will, / But in Gods mercy who choseth whome he will."[29]

Jacob's disguise, with its concomitant challenge to the divine, provides, I suggest, a scriptural context for one early modern writer's evaluation of the danger of wearing pelts. Writing in 1633, William Prynne took a stand against wearing animal "skins or likenesse," arguing that they marked and, significantly, produced postlapsarian human corruption: "What is this but to obliterate that most *glorious Image which God himselfe hath stamped on us,* to strip us of all our excellency, and to prove worse than bruits?"[30] Where Calvin's criticism of Rebekah is that she "obscures" the prophecy of God, Prynne's attack on the pelt wearer (an attack for which his ears were clipped like a sheep's) is that such dressing up "obliterates" the image of the divine.[31] For Prynne, it is what animal skin does to a human (and not what a human does to an animal) that is important, and what might be termed the thingness of the animal pelt can be found, as Brown suggested, in the change in the subject-object relations. Wearing skin alters the wearer: the animal-made-object becomes a thing asserting itself in the world, and the human becomes just a moldable body and not "the temple of God," as 1 Corinthians 3:16 has it.

Animal skin, for Prynne, is thus no longer simply the symbol that reminds us of who we are (Stubbes's position); it is a powerful, active thing with destructive potential. It is worth quoting a little more from Paul's first letter to the Corinthians to clarify Prynne's true anxiety here: "Know ye not that ye are the temple of God, and that the spirit of God dwelleth in you? If any man shall defile the temple of God, him shall God destroy; for the temple of God is holy, which temple ye are" (3:16–17). The ambiguity of "him shall God destroy" seems to fit Prynne's meaning perfectly. Wearing animal skins will lead God to destroy the individual because the individual has polluted God's supreme work on earth. But the phrase might also be read as implying that polluting the human body is an obliteration of God, who is the source of all order, and thus that dressing in animal skins is a destruction of order itself.

Animal skin is thus much more than simply a product and thence an illustration of human dominion. Indeed, I want to propose that the objectification of animals that is given form in leather cannot be taken as an iteration of human dominion at all, as it is that very objectification that produces the animal thing, which in turn changes subject-object relations in its revelation of the fragility of the human. Thinking about animal skin, as Laurie Shannon has noted, reveals human skin to be insufficient. In George Wither's terms, humans lack "nat'rall

Armour"; they are frail, unlike animals that are born with thick or protective skins.[32] It thus becomes necessary for humans to wear pelts, and in being necessary the nonhuman becomes dangerously potent; the animal-made-object becomes, in short, a thing. Without animals we are nothing, but with them we are not what we might hope to be.

Skin's power, though, should be regarded as both particular and general. It is particular in that *wearing* an animal-made-object is different from, say, *eating* one. In eating, as I have argued elsewhere, it is the taking in of the animal—its ingestion—that is most discussed in early modern writing.[33] But wearing skin also reflects a more general conception of animal things, in that the power that resides in the skin of an animal is a power that can also be found in other animal-made-objects, including meat. Thus I want to make a large claim: the persistent presence of the animal in the animal-made-object seems always to defy the objectification that attempts its absenting. Dismembering a living animal actually *produces* new agents. Thus, to return to the dog-skin gloves, you might say that the animal's doggedness is not simply a leftover of what has been (it is not just a residue of the living animal that has been removed in the production of leather). The doggedness is actually a *product* of the objectification itself. This is an assertion that I will illustrate by turning to look at another Renaissance animal-made-object: civet.

The Smell of a Human

"That which wee call Ciuet is nothing else but as it were a superfluous sweate found betweene the flanks of a beast much like vnto a Cat."[34] So wrote Thomas Johnson in 1595. A more accurate description comes from the historian Karl H. Dannenfeldt, who writes that civet "is a soft, fatty, yellowish glandular secretion formed in a discharging pocket of two sacs or perineal glands located between the anus and the genitals of both the male and female civet cat."[35]

Civet is also very much—in Europe, anyway—a Renaissance animal-made-object. For while, as Dannenfeldt notes, castoreum from beavers, musk from male deer, and ambergris from the stomach of the sperm whale were all known in the ancient world as sources of or foundations for perfume, civet was only discovered by Europeans in the mid-fifteenth century, and only became widely available as a commodity from the early sixteenth century onward, as colonial expansion into Africa and Asia took place. The value of the animal—like a sheep's value—lay in its continuing productivity: it was not killed when civet was removed, and this made the animal a profitable commodity.

15 *Civet Cat.* From Edward Topsell, *The Historie of Fovre-Footed Beastes* (London: William Jaggard, 1607), 757.

As with skin, civet both constructs and upsets notions of being human. Like leather, it is worn. But civet goes beyond leather. Not only is perfume put *onto* the human body, thus changing its external manifestation—its smell rather than its appearance in this instance. Perfume also enters *into* the body—as does meat, but here in the form of aroma. And it is in this doubleness, I think, that the specific power of civet lies. Civet has the potential to transform the human both from without and from within, and this is revealed in its impact on human smell—and I mean that in both internal and external senses: the aroma given off by the human, but also the human ability to sense the aroma of the world. In the first meaning of "human smell," civet does not simply remind humans that, like animals, they give off an odor. Wearing perfume (with its animal foundation) reveals that early modern humans actually *chose* to smell like animals, and thus that the human will, which should keep the human human, seemed to work against them. So, in 1622, in another kind of human will, John Jane, a sailor, was a man of his age when he bequeathed thirty shillings to one colleague on his ship the *Charles,* adding, "Alsoe I giue vnto John Betchered and Nicholas Johnson both belonging to the said shipp one Civet cat equally betweene them." Jane knew the value of his Renaissance animal-made-object.[36] Indeed, the early modern period is of particular note in relation to civet. Alain Corbin has found that in the late eighteenth century, "ambergris, civet and musk went out of fashion" for those higher up the social scale, that the Enlightenment brought with it a belief that animal scents "belonged to the masses."[37] Philip Stubbes, writing two hundred years before Corbin traced this shift, apparently was ahead of his time when he argued that "these . . . palpable odors, fumes, vapours, smells of these musks, cyuets, pomanders, perfumes balmes & suche like ascending to the braine, do . . . denigrate, darken and obscure y[e] spirit and sences, then either lighten them, or comfort them any manner of way."[38] The human senses are "obscured" by scents, just as Rebekah obscured the will of God through goat skin. And, to return to Introna's image, the "fumes" do their invisible work and challenge human intent.

In the second meaning of "human smell" (our capacity to sense the aroma of the world around us), humans experience another kind of loss of controlling power. As Eleanor Margolies puts it, "Smells pose a pungent challenge to philosophies of autonomous action. . . . [They] do not remain attached to their source, nor respect boundaries," but instead display a freedom that interrupts the perceived dominion of humans, thus revealing that dominion to be mythical and not real.[39] But it is not just its power to travel beyond the rule of humans that marks the danger of smell. In *Dialectic of Enlightenment* Horkheimer and Adorno argued that "when we see we remain what we are; but when we smell we are taken over by otherness."[40] Stable identity is shattered by odor, and perfume can thus be read as a self-inflicted wound to that superior being called the human. "I think, therefore I am," it seems, is challenged by "I smell, therefore I am something else." In these instances—through choice and through loss of control—civet not only reveals its own thingness; it also exposes the fact that the self can undo itself; that the will is recalcitrant; that humans themselves are things.

This linking of the role of the senses with a challenge to a humanist conception of human status echoes concerns raised in the Old Testament story of Jacob and Esau. Where earlier I read that tale as a source of one way of thinking about the wearing of animal skins in the Renaissance, I suggest that it also focuses on human perception and its poverty and provides a link between skin and civet—my two animal things—and the human. If Jacob's dressing in animal skins marks him as false, the "dim" eyes of his aged father, Isaac (who was once replaced by a ram), also construct him as dangerously lacking in that apparently human trait: judgment. Indeed, the trick Jacob plays on Isaac is an abuse not only of his father's blindness but of his other senses, too. Isaac's sense of smell is deceived because Jacob wears Esau's raiment, which Isaac sniffs to check the veracity of the body before him (as if clothing, unlike perfume, merely projected outward the natural aroma of the human individual within). Isaac's taste is fooled in that the meat Jacob brings him is not venison but kid and has actually been prepared by Rebekah and not by Jacob. It is, indeed, only Isaac's hearing that is true, but, ironically, Isaac does not trust it. Having touched his son, he says, "The voice is Jacob's voice, but the hands are the hands of Esau." After contemplation, Isaac trusts what was understood to be the lowest of the senses, touch.[41] His failing is not only that he thinks he can work against the will of God by blessing Esau (for God, after all, has already prophesied Jacob's rule over his older brother); it is also that he has faith in the tangible rather than the intangible. In this he echoes Esau, who gave up his birthright for a mess of pottage (Gen. 25:33). In 1592 Gervase Babington noted of that moment, "For what do they else then contemne spirituall things to obtayne earthly."[42] Skin, smell, stuff: all are linked to shatter the order of the human world.

The story of Isaac and Jacob is not the only place such linking of an animal thing and the failing human senses might have been encountered in early modern England. This dangerous lack experienced by the father who fails to see his true child in a story replete with references to the human senses reappears in a number of ways in Shakespeare's *King Lear,* a play that should be read as a repository for some negative contemplations of the nature of the human,[43] and that was, of course, written by the son of a glover.[44]

The subplot of *King Lear* contains a key parallel with Genesis 27. The letter that seems to reveal Edgar's parricidal desires that Gloucester is given by the apparently reluctant Edmund is false, like the skin Jacob presents to his father's touch at his mother's bidding. Gloucester asks, "You know the character [handwriting] to be your brother's?" Performing his role, Edmund hesitantly states, "If the matter were good, my Lord, I durst swear it were his; but in respect of that, I would fain think it were not." "It is his," Gloucester replies (1.2.60–64), echoing Isaac's judgment: "the hands are the hands of Esau." And the parallel is reinforced when we recall that "hand" in the early modern period was also, of course, a term for handwriting.[45] In both stories fraternal legibility and paternal lack of judgment are linked.

But the failure to tell the true child from the false is also, of course, the starting point for the main plot of *King Lear,* and for both Gloucester and Lear, their recognition of their failure to read their offspring leads to a focus not only on the senses but also on the status of the human more generally. Gloucester, on seeing Poor Tom, thinks "a man a worm" (4.1.33), and Lear also notes an alternative conception of his own species that uses a particular set of objects to make his point. When he catches sight of the disguised Edgar, Lear states: "Is man no more than this? Consider him well. Thou ow'st the worm no silk, the beast no hide, the sheep no wool, the cat no perfume. Ha! here's three on's are sophisticated; thou art the thing itself; unaccommodated man is no more but such a poor, bare, forked animal as thou art. Off, off, you lendings!" (3.4.100–106). This final sentence names clothes as temporary, as not owned, and for Jones and Stallybrass it presents clothing as a mark of "conflict" between (human and/or animal) source and wearer.[46] I, however, read this moment differently. To Lear, Edgar's apparent deficiency is both specific and general. Specifically, he is a human lacking in the use of the animal-made-object (silk, hide, wool, perfume). Because of this he is "unaccommodated"; he has no place.[47] I take this to mean that he has no place in either the kingdom or the chain of being. Humans, as Wither knew, need the animal-made-object to be who they hope they are. Without them, they are revealed as bare animals in borrowed pelts. With them, they are protected: "Robes and furr'd gowns hide all," Lear says in the following act (4.6.163). The animal-made-object, so he seems to believe, can mark and produce human power; and without such apparent markers the king sees all of humanity as uprooted. Lear, himself utterly displaced, of course, by his daughters, in fact links regal dominion with human dominion. And it is in the face of the breakdown of both of these structures that he tears off his clothes. Here, as if to underline the power of animal-made-objects—silk, hide, wool, civet—they are recalcitrant *even in their absence.*

In *King Lear,* then, clothes are no longer simply human possessions, unnoticed necessities (objects, silent workers). Their necessary-ness is their meaning, and this makes them things that, as Introna noted, makes them disturbing: "as we draw on them [perhaps we might adapt this to read, "as we draw them on"?], they become more and more part of who we are, or who we are becoming." Earlier, Lear had rebelled against denuding, claiming superfluity (2.4.262–65). But on the heath, in removing the "lendings," he acknowledges his true place as nothing more than "a poor, bare, forked animal," and a truly weak animal at that.[48]

But this acceptance is only temporary. And between the admission that man is "a poor, bare, forked animal" and the declaration that "A dog's obeyed in office" (4.6.157)—a belated and somewhat Latourian acknowledgment that power is embedded in a network rather than inherent in an individual—Lear interrupts his own thought process and makes a request that must be read as an attempt to return to the order that has been lost: "Give me an ounce of civet, good apothecary / To sweeten my imagination" (4.6.129–30). As Andrew Hadfield has shown, *King Lear* is a play full of medicinal herbs,[49] and here Lear believes—against Stubbes, but conventionally enough—in perfume's medicinal qualities. According to Edward Topsell, civet can "purge the braine,"[50] while a thirteenth-century Arab record presents it as "a good and useful remedy for faintness of the heart." But civet could also be taken orally: according to Pietro

Castelli, writing in 1638, when taken in wine, civet "cheered the heart."[51] All of these effects—purging the brain, remedying faintness of and cheering the heart—would be ideal for Shakespeare's aged, mad, and despairing monarch. But civet also has another power: it can, Lear thinks, return to him his ability to live in a cruel world: it can "sweeten [his] imagination."

In this call for civet, for the animal-made-object, Lear reveals that—despite his glimpse of another kind of human—he still clings to his faith in humanity's power over the natural world. He believes that his mental clarity (which is part of that natural dominion—John Donne terms such clarity a "disafforestation" of the mind)[52] can be reinstated by the presence of the animal-made-object and that he can, to return to Latour, "bind back" the hounds that he sees around (and within) him. There is a logic to this desire that returns us to the Bible. Genesis 2:19 states, "And out of the ground the Lord God formed every beast of the field, and every fowl of the air; and brought them unto Adam to see what he would call them: and whatsoever Adam called every living creature, that was the name thereof." The animals here are utterly passive: they are "brought before" Adam and are named. Lear wants to repeat this naming in what you might call worldly terms: he wants to see brought before him an animal-made-object that man has constructed. He wants to see human power as it is emblematized in the reduction of animal to product—the civet cat into the "ounce of civet"—to reassure himself that, at this moment of crisis, there is some order in the universe. He wants to return to a time when man was in complete control of meaning and was thus the sole possessor of agency: when men were men, you might say, and sheep were plants.

But, of course, this is a desire that can never be fulfilled. In his objectification of the civet cat into "an ounce of civet," Lear, like the sociologists Latour criticized, has assumed that an object can have no agency in itself, that the only changes that it can bring about are actually changes made by humans—by, in *King Lear,* the "good apothecary," who will mix up the medicinal simple.[53] But Lear's faith in the power of the human to transform its world is brought up short. In *King Lear* Shakespeare uses civet not as a cure but as an acknowledgment that a cure is beyond human capacity. He makes clear that possession of an animal-made-object cannot be taken simply as an enactment of dominion. The animal-made-object will not allow for that: it is a thing, and its thingness can be found in the paradox that Lear's request presents. The king wants, through an animal-made-object, to become human again, to become the dominant being in an ordered universe. But in this desire Lear reveals that his conception of the human depends on the persistent and recalcitrant presence of the animal thing, and thus is not the human he imagined at all.

This is something that is brought into violent focus in the play's final scene, when Lear—with awful irony—*howls* against what he regards as the disorderliness that makes no distinction between humans and animals. "Why should a dog, a horse, a rat, have life, / And thou no breath at all?" he asks (5.3.305–6). All his hopes—all his *humanist* hopes—are shattered. The hound has not been bound back to its cage; rather, it has run rampant. But Lear ends the play still pulling in two directions. On the one hand he asks for help to become the thing he has glimpsed that he is: "Pray you, undo this button: thank you, Sir" (5.3.308). This is naked man, helpless in the world. But, on the other hand, even as he asks that his clothing be removed, Lear is still clinging to something else: "Do you see this? Look on her, look, her lips, / Look there, look there!" (5.3.309–10). His dying hope is that there is, after all, a difference between Cordelia and a dog, a horse, and a rat. He is wrong. Subject-object relations that appeared stable are revealed to be fragile, and the final moments of the play are an acknowledgment not of human agency and superiority (evidenced in a king's power to divide his own kingdom—where the play began) but that the human is just one actor among many: that it is, indeed, "the thing itself." Or, to put it another way, in *King Lear,* a play that emphasizes the importance of the animal-made-object, the humanist ideal of the human as separate, as an end and not a means, is inevitably itself undone. "Break, heart; I prithee, break!" says Kent as the king dies (5.3.310), and we return, perhaps, to the world of dog-skin gloves and animal punning, for "break" is the correct verb for carving a hart—an aged male deer.[54] Thus Lear, at the end of it all, is just a dismembered mature male animal: he, like so many deer in England, is the commodity of a monarch, an animal-made-object.[55] He is a Renaissance animal thing.

Chapter THREE

The Cujo Effect

PAUL YOUNGQUIST

> And do not seek to know whether personally these gentlemen are in good or bad faith, whether personally they have good or bad intentions. Whether personally—that is, in the private conscience of Peter or Paul—they are or are not colonialists, because the essential thing is that their highly problematical subjective good faith is entirely irrelevant to the objective social implications of the evil work they perform as watchdogs of colonialism.
>
> —**AIMÉ CÉSAIRE**

> Dog people like to forget that dogs were lethal guided weapons and instruments of terror in the European conquest of the Americas.
>
> —**DONNA HARAWAY**

In Lewis Teague's 1983 horror film *Cujo,* a good dog goes bad.[1] A sweet, furry family pet acquires a taste for human blood—with ill effects. Unbeknownst to its owners, the fluffy St. Bernard has suffered a painful encounter with a vampire bat, usually native to the tropics but for cinematic purposes now resident in rural Maine. A bite to Cujo's nose awakens—or perhaps reawakens—the carnivore within. The wound won't clot, a clear sign of draculin, the glycoprotein in bat saliva that inhibits coagulation. Once infected, Cujo slowly goes ballistic, first turning against the abusive bumpkin who owned him, then against pretty much anybody who comes within reach.

Cujo is easy to dismiss as just another sleazy horror flick. But the line it draws—and crosses—between the animal as pet and as predator marks a significant historical and philosophical boundary. Because the film is so viscerally direct, it explores that boundary with uninhibited zest, raising the horrifying possibility that every doggie conceals a beast, that animal love is allied to terror. That's the discovery the protagonist makes when her car, a Ford Pinto with a bad carburetor, stalls in front of a perilously Gothic farmhouse where she's driven for repairs. Her small son sits in the other bucket seat sobbing in fear. Cujo has just mauled the car, and they're locked inside, marooned. They've met the dog before, on an earlier visit. Then he was hospitable, if a little testy. Now he's vicious and bedraggled. In fact, he's just eaten his former owner alive. Cujo has slipped the collar of domestic affection and lapsed into pure animal savagery. The little boy in the front seat is next on the menu, and he knows it. As the engine stutters and stalls, his mother asserts her terrified humanity: "Fuck you, dog. Fuck you."

Cujo may not achieve the heights of cinematic glory, but it's a strangely compelling, if compellingly confused, meditation on the theoretical and historical boundaries that demarcate the human and the animal. In this it bears comparing to a similar if less exuberant meditation, Giorgio Agamben's *The Open: Man and Animal.* In a reading of Heidegger that admittedly lacks the movie's intensity, Agamben describes what he calls "the anthropological machine," which is to say, "a device for producing the recognition of the human."[2] How do you know a human when you see one? It looks like an "anthropomorphous animal," an animal, that is, shaped like a man. The fundamental characteristic of a human, in other words, is that it looks like one—which could be said of a lot of animals, a duck, say, but not by those animals themselves. The constitutive difference that marks the *human* is that you *know* a human when you see one, or rather that a human *knows itself.* Agamben leans heavily on the strange nomenclature that Linnaeus uses to distinguish the human animal, the little loop of *sapiens* that gives *Homo* its human difference.

The human in this sense is an epistemological imperative. Agamben puts the point this way: "man is the animal that must recognize itself as human to be human"—which is to say that this animal also recognizes itself as animal, which is to say, as not man. *Homo sapiens* knows both that it is and is not animal, that it is and is not man, which means that we're pretty close here to a Hegelian flip into something higher. Agamben (or is it Linnaeus?) stops short of this flip, however, by holding man and animal in tension and requiring that the human recognize itself as (not) both. Obviously this tension puts a little strain on the notion of the human, since the anthropological machine, as Agamben later describes it, "functions by excluding as not (yet) human an already human being from itself, that is, by animalizing the human, by isolating the nonhuman within the human."[3] Thanks to this tension, the anthropological machine makes room for an *animal* man whose knuckle-dragging proximity disturbs the sweet self-awareness of the human. And this might lead, now and then, to the need to assert a boundary, not between man and animal but between the animal man and the human. What, one might wonder, would be the human cost of such a boundary? And how would you police it? Never much of a historicist, Agamben approaches such questions only indirectly in *The Open*. But for Teague they are the stuff of horror and, oddly, of history, too.

Cujo, after all, is a strange thing to call a dog. It conjures forgotten histories. This bloodthirsty St. Bernard shares a name with one of Jamaica's greatest Maroon warriors: the legendary Cudjoe. He and his fearsome bush fighters so harried British colonists and their militia that in 1739 Britain sued for peace and granted the Maroons, with certain stipulations, fifteen hundred acres of land and perfect liberty. The Maroons had been harassing the British almost from the moment the invaders snatched Jamaica from Spain in 1655, when many blacks slipped their shackles and either fought or fled into the woods and mountains of the tough interior. These blacks never bowed to British rule. The chaos of colonial conquest set them free.

This did not make them free men in the eyes of the British, but neither were they slaves. The Maroons inhabited a peculiar category somewhere between those poles, one that made them difficult to situate politically. Major General Robert Sedgwick, an early British commander, put the point with political if not quite grammatical precision: "Having no moral sense . . . and not understanding what the laws and customs of civil nations mean, we know not how to capitulate or treat with any of them."[4] The Maroons appear, according to General Sedgwick's description, to live beyond the pale of the human, wholly without the moral sense or civil authority necessary to secure peace through treaty. Even the word that names them suggests as much. A footnote to *The Proceedings of the Governor and Assembly of Jamaica, in Regard to the Maroon Negroes* (1796) gives two mutually reinforcing derivations for "Maroon." The first signifies "hog-hunter," since "*Marrano* is the Spanish word for a young pig"; the second more simply signifies "monkey," since in the French of the footnote, "maron" derives from "Simaran," a Spanish word for ape used to describe fugitive slaves.[5]

Either way, the Maroons as a group are distinctive for their proximity to the animal. What civilized nation would sue animals for peace? And yet after more than fifty years of harassment, forty-four acts of assembly to suppress it, and £240,000 expended, the British negotiated a "treaty of peace and friendship" with Cudjoe and his fierce fighting Maroons. The Maroons agreed to eradicate rebels, capture runaway slaves, and defend Jamaica in the event of invasion. In exchange, in the words of the treaty, "the said Captain Cudjoe, the rest of his captains, adherents, and men, shall be for ever hereafter in a perfect state of freedom and liberty."[6] Mirabile dictu, the treaty of 1739 appears to humanize the Maroons. Sedgwick's savages become something like citizens, capable of alienating a portion of their natural executive power to the collective judgment of a civilized community and its constitutive laws. The performative magic of juridical discourse assimilates them to civil society. Formerly too fierce to treat with, these animals become—well, *almost* human. To be precise, the Maroons come legally to inhabit Agamben's middle zone of animal man.

It's a tricky position. The language that legitimates this assimilation belongs to John Locke, whose *Two Treatises of Government* (1682) underwrites the aims and effects of the British treaty with Cudjoe. Much critical attention has been devoted recently to Locke's ambiguous attitude toward blacks and slavery.[7] As one of the authors of *Fundamental Constitutions of Carolina* (and heavily invested in the slave-trading Royal African Company), he wrote that "every freeman of Carolina shall have absolute power and authority over his Negro slaves, of what opinion or religion whatever."[8] His famous "just-war" defense, in the *Two Treatises,* of slavery as an institution, however, applies only to England and not her colonial empire.[9] In the words of one commentator, "Locke's theoretical and colonial writings do not cohere as a unified body."[10]

What does link the theoretical to the colonial in Locke is the political function not of the slave but of the animal. Locke opens a middle zone that resembles Agamben's, where humanity lapses into animal man. The Maroons inhabit this peculiar zone by virtue of an ambiguous humanity: they're human enough to treat with but remain animal enough to track and kill. They make good watchdogs, with this difference: they consent to their deployment. The civil standing of the Maroons thus remains peculiarly qualified. Locke is quick to insist that not all contracts produce a commonwealth, "only this one of agreeing together mutually to enter one Community."[11] The British treaty with the Maroons produced a community *apart,* a space of "freedom and liberty" whose purpose was to protect colonial society against insurrection and loss of property. The Maroons were hardly full participants in civil society. But neither were they beyond the touch of its ruling principle of law.

They played a kind of hokey-pokey with civil society, one foot in and one foot out. Although they realized few of the benefits of community that Locke describes (chief among them the accumulation of property), they remained subject to its rule of law and the consequences of its breach—but subject in a peculiar way that reflects their peculiar social position. Civil law for the Maroons, the primary social function of which is to secure the community against violence, coincides with natural law, the imperative to preserve human life. Any civil offense thus constitutes a crime not merely against the community but against nature. A breach of treaty therefore constitutes a double violation for the Maroons: it violates civil law but also natural law, which authorizes punishment and reparation for violating the imperative to preserve human life. The Maroon as animal man becomes doubly dangerous to civil society because a treaty violation constitutes a crime against both civil and natural order.

When this happens, the man succumbs to the animal. In Locke's words, the offender "becomes dangerous to Mankind, the tye, which is to secure them from injury and violence, being slighted and broken by him. Which being a trespass against the whole Species, and Peace and Safety of it, provided for the Law of Nature, every man upon this score, by the Rights he hath to preserve Mankind in general, may restrain, or where it is necessary, destroy things noxious to them." Under circumstances of trespass against the species, then, "a Man so far becomes degenerate, and declares himself to quit the principles of Human Nature and to be a noxious Creature."[12]

In effect, such a man declares war on the human community, vindicating the use of force against him, even unto death: "And one may destroy a Man who makes War upon him, or has discovered an Enmity to his being, for the same Reason, that he may kill a *Wolf* or a *Lyon,* because such Men are not under the ties of the Common Law of Reason, have no other Rule, but that of Force and Violence, and so may be treated as Beasts of Prey, those dangerous and noxious Creatures, that will be sure to destroy him, whenever he falls into their Power."[13] Here is where the colonial and the theoretical meet in Locke not in slavery per se but in the legitimation of force and violence against the lapsed animal man. Locke turns Agamben's anthropological machine up full throttle, "animalizing the human, by isolating the nonhuman within the human." Within the human lurks the animal man, and within the animal man lurks a wolf, whose force and violence, when insurgent, warrants its extermination. Such is the antipathy of civil society toward the noxious animal. Perish Maroons who rebel. "Fuck you, dog."

In 1795, after more than fifty years of keeping peace and maintaining prosperity, the Maroons declared war against the British. They had stifled many a rebellion and returned many a fugitive slave, living peaceably and independently, abiding strictly by the terms of the 1739 treaty. But by the 1790s they were growing restive, at least in Trelawney Town. The land that had once secured their freedom was hemming in their increasing numbers. The new British guardian who lived among them lacked the courage and élan of their old favorite. Then there was the incident of the pig, stolen in Montego Bay by two dissolute Maroons who were punished, in violation of the treaty, by the British, who used a slave to wield the whip.[14] The insult pushed the Trelawney Town Maroons to the breaking point, and they commenced hostilities.

The lieutenant governor of Jamaica, the Earl of Balcarres, declared martial law and mobilized local militia, dragoons, and infantry—several thousand soldiers in all, but it was hardly a fair fight.[15] Having fewer than five hundred men at arms, the Maroons were nevertheless a wily enemy, expert bush fighters who refused to meet the British on the open field. Instead they drew His Majesty's troops into blind mountain corridors and rocky cockpits, hiding in ambush and then suddenly splattering them to bits with an invisible hail of lead. R. C. Dallas, the contemporary historian of the Maroon War, conveys a grudging respect for rebel methods: "A small body of negroes defied the choicest troops of one of the greatest nations in the world, kept an extensive country in alarm, and were at length brought to surrender, only by means of a subvention still more extraordinary than their own mode of warfare."[16] Hard fights call for hard measures, particularly when the enemy abdicates the man and declares war against the human community. Nor was this offense against the species a local phenomenon. In St. Domingo, 260 nautical miles away, Dallas tells us, "the Negroes had been driven into arms by the French Government, in support of doctrines unfounded in nature, and peculiarly hostile and destructive to the order and well-being of every West Indian colony" (1:142).[17] Should those doctrines have found their way to Jamaica, the Maroons might have proved an inspiring example for a subjected population of blacks who outnumbered their white masters ten to one, and, as Dallas observes, "it is highly probable that the example might in time have united all the turbulent spirits among the slaves in a similar experiment, if not in the same interest" (2:2). Under such conditions, "the lives of the colonists must have been spent in continual terror; massacre and depredation would have spread throughout the country, and all the credit of the island in Great Britain would have sunk to nothing" (2:2–3). Time to take off the gloves and deal with the Maroons like the noxious creatures they had become.

The Assembly of Jamaica decided to call in "auxiliaries." That's the word Dallas uses to describe the weapon of choice and duress deployed against the bush fighters, a supplementary weapon for a supplementary antagonist, animal man turned noxious pest. "It occurred to Colonel Quarrell," Dallas writes, "that the assistance of a certain number of the Cuba chasseurs would be attended with happy effects: he foresaw that the very terror they would spread would induce the Maroons to submit on proper terms; and he argued, that even if the commander in Chief were compelled to bring them into actual service, it would be better, and more for the interest of humanity, that some of the rebels should be

16 *A Spanish Chasseur of the Island of Cuba,* 1803. Engraving. In R. C. Dallas, *A History of the Maroons from Their Origin to the Establishment of Their Chief Tribe at Sierra Leone,* 2 vols. (London: T. M. Longman and O. Rees, 1803), vol. 2, frontispiece.

thus destroyed, than that the most barbarous massacres should be committed on the inhabitants, and the colony ruined" (2:6).

In the service of those happy effects, Quarrell procured from Cuba, with not a little skullduggery, about forty Spanish handlers known as "chasseurs," that is, hunters, and one hundred dogs—bloodhounds (fig. 16). "These auxiliaries," Dallas writes, "were more formidable than the finest regiment of the most warlike nation could have been, and from the time of their being employed, neither surprise nor ambush annoyed the troops" (2:5). Their reputation preceded them to Jamaica, as Bryan Edwards notes with literary flair: "If entire credit had been given to the description that was transmitted through the country of this extraordinary animal, it might have been supposed that the Spaniards had obtained the ancient and genuine breed of Cerberus himself, the many-headed monster that guarded the infernal regions."[18] But no. The Spanish bloodhounds were living animals bred to vicious purpose. Dallas's description is more to the military point: "The animal is the size of a very large hound, with ears erect, which are usually cropped at the points; the nose more pointed, but widening very much towards the after-part of the jaw. His coat, or skin, is much harder than that of most dogs, and so must be the whole structure of the body, as the severe beatings he undergoes in training would kill any other species of dog" (2:57–58).

These bloodhounds are biological weapons deployed against an enemy whose animal ferocity justifies a response in kind.[19] They hunt the animal man to secure the prosperity of the human community. Blood, or at least the scent of it, lubricates the anthropological machine. Upon arrival in Jamaica, bloodhound and chasseur made an immediate impression on the populace: "The muzzled dogs with the heavy rattling chains ferociously making at every object, and forcibly dragging on the chasseurs, who could hardly restrain them, presented a scene of a tremendous nature, well calculated to give a most awful colouring to the report which would be conveyed to the Maroons" (2:120). These were terrifying animals. And while Dallas does his best throughout his narrative to persuade the public that they remained always in their handlers' control, he lets one slip the leash in a footnote, illustrating their ferocity:

> It is hardly worth while to mention an accident by which an old woman lost her life, but it has been suggested that the omission of it may receive an unfavourable construction. One of the dogs that had been unmuzzled to drink when there was not the least appprehension [*sic*] of any mischief, went up to the woman, who was sitting attending to a pot in which she was preparing a mess. The dog smelled at it, and it was troublesome; this provoked her, she took up a stick and began to beat him, on which he seized on her throat, which he would not let go till his head was severed from his body by his master. The wind-pipe of the woman being much torn, she could not be saved." (2:169n)

The bloodhound scents, seizes, and holds on to the very death—its own as well as its prey's. It is a guided missile that explodes on impact, slowly and sadistically.

This is why Marcus Rainsford, a captain of the Third West India Regiment and eyewitness to the deployment of bloodhounds against the Maroons, called their use "the savage custom of a barbarous age (only employed exclusively against the worst criminals)."[20] The dogs of war fight savagery with savagery. And how exactly does this biological weapon get built? Here, too, Rainsford proved the better historian.

> From the time of their being taken from the dam, they were confined in a sort of kennel, or cage, where they were but sparingly fed upon small quantities of the blood of different animals. As they approached maturity, their keepers procured a figure roughly formed as a negro in wicker work, in the body of which were contained the blood and entrails of beasts. This was exhibited before an upper part of the cage, and the food occasionally exposed as a temptation, which attracted the attention of the dogs to it as a source of the food they wanted. This was repeated often, so that the animals with redoubled ferocity struggled against their confinement while in proportion to their impatience the figure was brought nearer, though yet out of their reach, and their food decreased, till, at the last extremity of desperation, the keeper resigned the figure, well charged with the nauseous food before described, to their wishes. While they gorged themselves with the dreadful meat, he and his colleagues caressed and encouraged them. By these means the whites ingratiated themselves so much with the animals, as to produce an effect directly opposite to that perceivable in them towards the black figure; and, when they were employed in the pursuit for which they were intended, afforded the protection so necessary to their employers.[21]

Such brutal practices keep the anthropological machine running. Canine carnivores acquire a taste for raw meat in human form—black meat served up by white men. These ferocious bloodhounds scent, see, and secure the difference that distinguishes their keepers from the animalized men they hunt. It's a difference of race. Bloodhounds bestow blackness on the bodies they devour. Rainsford makes it clear—pace Dallas—that the chasseurs raised these dogs for exactly such savage ends: "The common use of them in the Spanish islands was in chace of runaway negroes in the mountains. Once they got scent of the object, they immediately hunted him down, . . . and instantly devoured him."[22] These dogs were bred to eat black slaves. They draw a line between white and black in blood.

It was as a weapon against savagery that the Assembly of Jamaica deployed the Cuban bloodhounds. To justify their use, that august body had first fully to animalize the animal man they acknowledged as such by treaty. A kind of advertising campaign accompanied the decision to use bloodhounds that recalls in its finesse the depiction of Willie Horton in the American presidential campaign of 1988. The organizing genius of this campaign was Bryan Edwards, planter and famed historian of the West Indies, to whom it fell to write an introductory statement to *The Proceedings of the Governor and Assembly of Jamaica, in Regard to the Maroon Negroes*. As the official publication of documents pertaining to the Maroon War, the pamphlet justifies state-sponsored savagery in the name of defeating savages.

Following Edward Long's *History of Jamaica* (1774), Edwards describes the Maroons as living "a savage way of life" from their earliest beginnings. They exist "wholly without any reflection on the past, or solicitude for the future." They perform their contractual police function with brutal zeal, manifesting "a bloodthirstiness of disposition, which is otherwise unaccountable." They put down rebels with wicked verve, in some cases taking the head for a trophy or eating the heart in triumph. Their bodies may be beautiful, but they prefer their beef putrid. Woe to the whites who fall into their sanguinary hands. They are "murdered in cold blood, without any distinction of sex, or regard to age . . . and the shrieks of the miserable victims, which were distinctly heard at the posts of the British detachments, frequently conveyed the first notice that the Maroons were in the neighbourhood." In short, the Maroons were "a banditti of assassins," "an army of savages." "Tenderness toward such an enemy," Edwards concludes, "was cruelty to the rest of the community."[23]

There are, of course, other ways to construe the cruelty involved in calling out the dogs. Edwards predicted that "urging the canine species to the pursuit of human beings, would probably give rise to much observation and animadversion in the mother country," and he was right.[24] Parliament debated the issue in March 1796, when, the London *Times* reported, one General Mcleod "proposed a bill to ban the use of bloodhounds against the Maroons in Jamaica. He reminded his peers of the Maroons' honorific past: 'They were descendants of those whose freedom our ancestors had acknowledged.'" Mcleod impressed on the House of Commons "how contrary to all the dictates of honour and spirit this mode of warfare was," comparing British deployment of these savage weapons to the barbarities of the Spanish in conquering the New World: "Thus would Great Britain have the infamous honor of pre-eminence in cruelty." Henry Dundas carried the day, however, insisting of the dogs "that they were not used for the purpose of tearing the Negoes to pieces, but merely to prevent run-away Negroes from escaping."[25] No cruelty there. As we all know, deterrence does not equal destruction. Apparently the question of cruelty cannot reasonably be asked in the absence of bodily suffering. So long as the Maroons were not eaten alive, the deployment of dogs against them remained justifiable.

Something almost magical occurs in this formulation, however. It extenuates the practice of savagery by simulating its effects. If no Negroes are torn to pieces, no cruelty can have occurred. Because the terror of that prospect produces the same end, the civilized deployment of savage weaponry aims not to destroy but simply to terrify. Terror in this sense becomes a sign of animal savagery, evidence of the animalization of the human that legitimizes destruction. This strange logic, by which fear of animal savagery becomes a sign of animality, explains one of the consistent effects of the deployment of bloodhounds against the Maroons. The terror the Maroons feel marks them racially. It becomes a sign of their blackness, evidence of racial difference. Those savage dogs patrol borders—black/white, Maroon/British, animality/humanity—not to secure but to produce them, as descriptions of black reaction to those bloody hounds confirm. Edwards attests that the bloodhounds "had certainly a powerful and very salutary effect on the fears of the rebel Maroons, a large party of whom now displayed strong and indubitable evidences of terror, humiliation, and submission."[26] Dallas claims placidly that "the activity of the chasseurs no negro on earth can elude" (2:61). And Rainsford, always most alert to the motivations behind such claims, notes that the whole point of deploying these biological weapons in the field was "to render them more terrific to the blacks."[27] The military purpose of bloodhounds is to produce terror, and the political effect of that terror is to produce race. The anthropological machine has a terror attachment that racializes as it animalizes humanity.

Dallas describes the Maroon surrender as that beautiful military achievement: a "bloodless triumph" (2:170). The arrival of the bloodhounds turned the tide of the war. General Walpole, leader of the British forces, Dallas tells us, "conceived the wisest step he could take would be to seize the opportunity offered by the terror which was spread by the arrival of chasseurs" (2:132). Seize it he did. While Walpole was careful to keep the dogs and their handlers to the rear of his troops in order to avoid mayhem, the terror they inspired everywhere preceded their formidable appearance. Of the chasseurs Dallas relates that "wherever they went, terror flew before them, anticipating their operations" (2:150). Imagine being hunted by dogs trained to seize, maim, and devour. Imagine becoming Maroon. It wasn't long before "the Maroons, terrified by the thoughts of the chasseurs, had resolved upon peace and submission" (2:148).

By late December 1795, many were ready to capitulate. The victory belonged to the dogs. "Thus concluded hostilities," writes Dallas, "without recourse being once had to the assistance of the chasseurs, beyond the operation of the terror they inspired" (2:168). Terror won the Maroon War, the deployment of the fear of savage dogs as much as the dogs themselves. So concluded the Assembly of Jamaica in its joint statement to Lord Balcarres: "We cannot but take this opportunity of

17 Marcus Rainsford, *The Mode of Training Blood Hounds in St. Domingo, and of Exercising Them by Chasseurs*, 1805. Engraving. In Rainsford, *An Historical Account of the Black Empire of Hayti* (London: Cundee, 1805), tipped in between 388 and 389.

expressing our acknowledgements of the eminent advantages derived by the importation of the chasseurs and dogs, in compliance with the general wishes of the island. . . . We are happy to have it in our power to say, that terror excited by the appearance of the dogs has been sufficient to produce so fortunate an event; and we cannot but highly approve that attention to humanity so strongly proved by their being ordered in the rear of the army."[28]

The deployment of bloodhounds, it turns out, becomes an act of humanity through the terror it produces in savages who menace civil society. If that terror is a sign of racialized savagery, its production reasserts the human, since it provides yet another occasion for humanity to recognize itself as (not) animal. But dogs are difficult co-pilots of the anthropological machine. Their bite can be worse than their bark, collapsing a politically conducive terror back into simple cruelty. As usual, Rainsford appreciates the high risk taken by the British in Jamaica: "Though a successful, yet it was a dangerous experiment, and one which will, it is hoped, never be again tried by British soldiers."[29] Humanity hangs in the balance.

18 Marcus Rainsford, *Blood Hounds Attacking a Black Family in the Woods,* 1805. Engraving. In Rainsford, *An Historical Account of the Black Empire of Hayti,* tipped in between 422 and 423.

And not only on the battlefield. Military weapons have domestic effects. Sometimes they misfire, targeting the innocent. Dallas's account of the old woman's slaughter is a case in point. That the rogue bloodhound attacked her while she was cooking shows that terror has a domestic register. It targets relations of home and hearth, albeit indirectly. Dallas brushes the incident off as an anomaly, but it is part of a pattern of collateral damage that haunts the deployment of bloodhounds in Jamaica. During the debate about them in London, Courtenay, a member of Parliament, worried about whether the chasseurs could in fact restrain their dogs "from executing the cravings of their savage instincts," and provided a grim reminder of their ferocity: "When they were first landed on the beach at Jamaica, they seized on a soldier's wife with a child in her arms, and they were bayoneted before they could be compelled to quit their prey."[30] Here, dogs literally rip a family apart. Rainsford provides the clearest illustrations of such domestic violence. Two are visual. An image of the wicker figure used to train bloodhounds to eat blacks shows it to be female (fig. 17). Another image is even more horrific because less military: dogs attack a woman with a baby in her arms, their savagery doubled by a second attack against a family in the background (fig. 18). These images force the conclusion that targeting rebels targets families, too, or, more specifically, mothers and children, dependents in the little sovereignty of marriage. The terror of bloodhounds on the battlefield proves useful in the home. Its racial savagery reinforces dependence.

Rainsford puts the point in its full horror when describing the French deployment of dogs in St. Domingo, until even he, a seasoned soldier, can take no more:

> The dogs frequently broke loose in the vicinity of the Cape, and infants were devoured in an instant from the public way! At other times they proceeded to the neighbouring woods, and surprising an harmless family of laborers at their simple meal, tore the babe from the breast of its mother, or involved the whole party, and returned with their horrid jaws drenched in the gore of those who were acknowledged, even in the eyes of the French army, as innocent, and therefore permitted to furnish them with the produce of their labor. Huts were broken into by them, and *
> The picture becomes too dreadful for description even for the best of purposes.[31]

The slaughter of innocents is for Rainsford no innocent by-product of unruly weaponry. Bloodhounds target mothers. They turn infants into fodder. In a manner so horrifying it elicits self-censorship, they invade the home, making it a scene of unspeakable horrors. Rainsford's lapse into asterisks differentiates the human from both the savage animal and its prey. This is the domestic correlative to the "humane" military deployment of bloodhounds: women and children as collateral damage. Their vulnerability reinforces dependency, tainting them, too, with a touch of the animal. Bloodhounds, it turns out, produce effects far beyond the battlefield. The terror they inspire proves socially useful. It reinforces racial difference and domestic dependency. When the British deployed weaponized dogs against the Maroons, they set a military and civil precedent that would prove perennially useful to civil society.

Coda: Cujo's Brood

Dogs remain military weapons. In the words of U.S. Army Regulation 190-12 (1993), "Trained dogs have been used by most of the world's military forces

19 Sgt. Michael Smith with his dog, Marco, watching a detainee on an unspecified date in 2003 at the Abu Ghraib prison in Baghdad, Iraq. © 2011 Associated Press.

since the first military units were organized. From these ancient beginnings the Military Working Dog's training has been continuously refined to produce a highly sophisticated and versatile extension of the soldier's own senses."[32] In the language of a more recent version of this publication (2007), dogs and their handlers are "force multipliers" (10). Their use has been refined since bloodhounds were deployed against the Maroons. Dogs and handlers now work in teams, one dog to one handler. The stated mission of the MWD is "to enhance the detection capabilities of the combat support force, and to provide psychological deterrent to hostile intrusions" (1993, 15). Critical to this mission is "controlled aggression," the capacity of the MWD to attack on command, bite, and hold for at least fifteen seconds (1993, 28).

Controlled aggression is a use of force, and as per policy set forth in Army Regulation 190-14, force must be finely calibrated to circumstances: "personnel will use the minimum amount of force to reach their objective."[33] The force of an MWD is greater than that of a club but less than that of a gun, "because a patrol dog is trained to terminate an attack on voice command of its handler" (2007, 11). The primary mission of the MWD today is not to induce terror. It is to multiply force, indirectly through its keen senses and directly through its disciplined aggression. The MWD nevertheless advances the function of the anthropological machine, situating the object of its aggression in the strange no-man's-land of animalized humanity. AR 190-12 (2007) admits as much by prohibiting the use of MWDs to produce terror, at least in a controlled environment: "Military working dogs . . . will not be used as part of an interrogation approach nor to harass, intimidate, threaten, or coerce a detainee for interrogation purposes" (12). Military Working Dogs, after all, can terrify.

That last caveat was of course added after the infamous events at Abu Ghraib prison in 2003 and 2004, where Arab detainees were abused and tortured in extremely humiliating ways, including the use of Military Working Dogs to induce terror during interrogation (see fig. 19). Perhaps such episodes are less aberrant than we are told to believe, less the bad behavior of a few rogue interrogators than the effect of that terror attachment wired up to the anthropological machine that produces and polices racial difference in civil society. How exactly did MWDs come snarling and snapping into that Iraqi prison? By way, strangely, of Cuba, a fact that brings to mind the Maroon War. On a visit to Abu Ghraib, Major General Geoffrey D. Miller, in charge of the prison at Guantanamo Bay in 2003, suggested toughening interrogation tactics, for instance, by using dogs to terrorize detainees. The grisly and familiar effects of his suggestions have been called "Animal House on the Night Shift," hardly a flattering phrase to animals but symptomatic of the operation of the anthropological machine.[34] Even more arresting, however, is the wording of Lieutenant General Ricardo S. Sanchez's memorandum of 14 September 2003 describing the amped-up methods of interrogation being authorized for Abu Ghraib. Under the heading "Presence of Military Working Dogs" runs the following revealing sentence: "Exploit Arab fear of dogs while maintaining security during interrogations."[35] The tactical deployment of terror against a recalcitrant adversary should sound chillingly familiar. Never mind that in this case the adversary had already been incarcerated. What matters is the deployment of dogs to produce the difference, in this case the very Arab difference, of the animalized human. That the Military Working Dog produces the terror it provokes is a subtlety lost on Sanchez. His memorandum backloads terror into ethnicity, racializing the detainee. It defines the Arab as the animal man incapable of recognizing the animal for what it—and he—is not. The anthropological machine continues to run at military speed for as far as the American eye can see.

Agamben says it is idling.[36]
I wonder how to shut it off.

Chapter FOUR

On Vulnerability

Studies from Life That Ought Not to Be Copied

RON BROGLIO

In his 1806 encyclopedia of images, *Microcosm,* William Henry Pyne illustrates the daily tasks of rural peasant life. Workers draw water from a well, bale hay, shoe horses, tend cattle, and watch over sheep. As A. E. Santaniello's introduction explains, "The men and women depicted in *Microcosm,* those masses of English workers, led hard lives; what was hardest was that they were also silent and, to most of their more fortunate compatriots, invisible. *Microcosm* was one of the first books to do for the common man the great and lasting service of making him visible."[1] The images are set out according to typical picturesque groupings and so provide professional and amateur artists with "studies from life" from which to take inspiration. Within the two-volume work, one page stands out as a unique set of images, titled *Slaughter-houses* (fig. 20). While illustrating the slaughter of cattle by pole ax, the artist has cleverly hidden the deadly head of the ax. We never see the blunt mallet head of the instrument.[2] This omission haunts the moment of death. The textual commentary by Pyne's companion C. Gray informs the reader that this plate of images "ought not to be copied," and so Gray underscores the omission where the instrument of death should be. Gray insists that the animals should not be copied despite the observation that "slaughtering cattle is a necessary evil."[3]

I am not interested here in the obvious animal rights claim that empire builds itself on the flesh of animals. True as this may be, there is a more complex series of connections between the working class and the evasive depiction of animal death. This dynamic of the nonrepresentable event housed within the bucolic can be clarified through an examination of the pastoral mechanisms that increase the productivity of peasants while circumscribing laborers as political subjects. The machinery of biopower at work on the material, biological, and political body of the agricultural laborer can be read through the figure of animal death. More potently, in the agricultural revolution, human life and labor function in the same manner as the life and labor of livestock. Human life and animal life are interconnected.

This chapter examines the representation of the slaughter and death of livestock in peasant poetry and the picturesque. During the eighteenth and early nineteenth centuries in Britain, landscape poetry and painting that self-consciously depict the countryside often obfuscate the end result of farmhands' labors. My conjecture is that the oddities and omissions surrounding representations of livestock death arise from discomfort over the mutual vulnerability felt between humans and animals. Cary Wolfe describes this as "the physical exposure to vulnerability and mortality that we suffer because we, like animals, are embodied beings."[4] It is a vulnerability felt particularly by the peasants. Indeed, the embodiment of these "embodied beings" entails the friction of contact between humans and animals in ways not common to the upper classes. Animal death reminds the worker of his or her own expendability. The body of the laborer and that of the animal are often given value not in and for themselves but rather as means toward serving national prosperity. Their embodied being and daily tasks lead not to their own benefit but to the appropriation by others of the value of their work. The death of domesticated animals is of interest because, like these animals, laborers lived within a social structure that they served but that rarely looked out for their welfare.

A word of caution before proceeding. Equating laborers with animals is too facile an analogy. It is true that, given their low social standing, rural laborers may well have felt their lives to be close to that of domesticated animals. Indeed, the comparison was often made, as in Timothy Nourse's *Campania Foelix, or a Discourse of the Benefits and Improvements of Husbandry* (1700), in which Nourse writes, "as with the men, *so are the Cattle.*"[5] Historical specifics of the agricultural revolution and the technologies and techniques devised in this period provide a clearer understanding of the animal-laborer dynamic. Of interest here are the specific forces in the eighteenth and early nineteenth centuries that made vulnerability viscerally potent for laborers. This chapter begins with a series of representations of livestock death and the discomfort such death prompted. It then examines how animals and humans fit within what was termed "improved agriculture." With the rise of statistics and land surveys, and with the enclosure

20 J. Hill, *Slaughter-houses,* c. 1806. Aquatint. From William Henry Pyne, *Microcosm, or A Picturesque Delineation of the Arts, Agriculture, and Manufactures of Great Britain,* 2 vols. (London: William Miller, 1806; reprint, New York: Benjamin Blom, 1971), 2:259.

movement shifting the rhetoric of land toward the wealth of the nation, the state began to govern life, and to do so in a way that flattened distinctions between rural laborers and domesticated animals, at times valuing livestock as a greater benefit to the nation than the working poor.

Before detailing the relationship between animals and laborers, let us return to the opening example of Pyne's image and Gray's textual commentary. Gray insists that the deaths are unsightly, and he asks the reader to avert his gaze: "But let us turn from a scene which the artist has given us too lively a copy of, not to inspire us with horror, and which can only set the mind of sensibility a thinking of the hourly, indeed, the unintermitting scenes of

The increase in number of cattle and sheep, sold in Smithfield, will appear from lowing statement.

	Cattle.	Sheep.
1731 to 1740, yearly average	83,906	564,650
1741 to 1749	74,194	559,891
1750 to 1758	75,351	623,091
1759 to 1767	83,432	615,328
1768 to 1776	89,362	627,805
1777 to 1785	99,285	687,588
1786 to 1794	101,075	707,456

From this table the increase in number, from 1731 to 1794, appears to be about one but combined with the increase in weight, it will be as 3 to 1.

Fig. 1. Slaughterman haltering an ox previous to his being killed. 2. Knockin

21 Pyne, *Microcosm,* 2:18.

misery, and we fear, too often of cruelty, caused by our enjoyments at table of harmless animals. Necessary they may in part be; but the less we see and think of them, the better."[6] Despite such insistence, Gray provides a table giving the weight of animal carcasses sold at London's Smithfield market from 1710 to 1804 (fig. 21). The average weight of a cattle carcass rose from 370 to 800 pounds, sheep from 28 to 80 pounds, and lambs from 18 to 50 pounds during this period. Gray goes on to list the number of cattle and sheep sold at Smithfield every decade from 1731 to 1794. Here, too, numbers increase, starting with a yearly average of 83,906 cattle and 564,650 sheep in 1731, and ending with 101,075 cattle and 707,456 sheep in 1794. At the level of statistical averages, Gray has no problem drawing dead animals. He provides a "cruelty, caused by our enjoyments at table," but in this case his tables are numerical and sufficiently abstract from the flesh on dinner tables to meet his approval. Indeed, as will become evident in the discussion of John Sinclair and the Board of Agriculture, below, statistics drove the demand for more meat.[7]

Gray is not the only one to avert his eyes from slaughter. In *The Farmer's Boy* (1800), the peasant poet Robert Bloomfield paints a bucolic picture of rustic labor throughout the seasons. Amid praise for labor and the liberty of English life, Bloomfield intertwines the life of animals, both domestic and wild. In putting his plow horses to bed, he "pats the jolly sides of those he loves." He milks the "full-charg'd udders" of the cows and "tugs o'er his pail, and chants with equal glee."[8] The farmhand Giles lives in close proximity to animals, and in concert with them labors to make the land productive. Such contact makes the spring slaughter of young lambs difficult:

> Ah, fallen rose! sad emblem of their doom;
> Frail as thyself, they perish while they bloom!
> Though unoffending innocence may plead,
> Though frantic ewes may mourn the savage deed,
> Their shepherd comes, a messenger of blood,
> And drives them bleating from their sports and food.
> Care loads his brow, and pity wrings his heart,
> For lo, the murd'ring BUTCHER with his cart
> Demands the firstlings of his flock to die,
> And makes a sport of life and liberty! (22)

The shepherd who identifies his labors with English liberty and, metonymically, with the joy of the young lambs has briefly revealed the darker side of the English pastoral landscape. Pointing to this passage, Donna Landry claims that "the georgic ethos rewrites the pastoral as fantasy, and itself as pragmatic reality, but it cannot exist without feeding on the very pastoral it repudiates."[9] Giles's liberty comes at the expense of other life. This realization is too much for the poet, who ends the section with patriotic phrases designed to cover the omitted scene of the animals' death:

> Down, indignation! hence, ideas foul!
> Away the shocking image from my soul!
> Let kindlier visitants attend my way,
> Beneath approaching *Summer's* fervid ray;
> Nor thankless glooms obtrude, nor cares annoy,
> Whilst the sweet theme is *universal joy.* (23)

Once the butcher's deed is done, the poet celebrates the land's bounty, epitomized in the landowner's feast with his workers. Old and young, day laborers, yeomen, and artisans join the feast, where "plenty reigns" and every food is to be had that "made our great forefathers brave." But what of the lamb or the pigs and chickens featured earlier in the poem? What of the cattle? Again, the animal's death is omitted since it is an unpastoral but "necessary evil," inscribed within the feast of all that is good in England's green and pleasant land. In the feast, the mutual vulnerability Giles felt with the lambs—whose death "makes a sport of life and liberty!"—becomes a thankful celebration for having survived where others were killed. The farmhand's work is hard, but at least he has not died for it.

After stanzas of praise for the harvest table, Bloomfield changes the table arrangements. These feasts where "The master, servant, and the merry guest, / Are equal all" have changed (45). Those *were* the days:

> When Pride gave place to mirth without a sting;
> Ere tyrant customs strength sufficient bore
> To violate the feelings of the poor;
> To leave them distanc'd in the mad'ning race,
> Where'er Refinement shews its hated face:
> Nor causeless hated; . . . 'tis the peasant's curse,
> That hourly makes his wretched station worse;
> Destroys life's intercourse; the social plan
> That rank to rank cements, as man to man:
> Wealth flows around him, fashion lordly reigns;
> Yet poverty is his, and mental pains. (46)

Labor is not for shared benefit; rather, it is divided with a "separate table and the costly bowl" (48). It seems that it is not only the butcher who "makes a sport of life and liberty" but also the landowner, who has separated himself from those who produce the bounty of his fields (22). Bloomfield ends the scene and this section of the poem with the lines "Let labour have its due; . . . then peace is mine, / And never, never shall my heart repine" (49). One wonders what happens if labor is not given its due. Certainly, the peace of the peasant would be disturbed, but would the master's peace be threatened as well? Lingering in these lines is the latent possibility of more spilled blood, not animal blood but human.

As a boy, Bloomfield had worked as a day laborer on his uncle's farm in the village of Honington, Suffolk. He was too small and frail to work in the fields and so did other, more menial tasks. Eventually, he left farm work to apprentice under his brother as a shoemaker in London, where he crafted his poetry about country life. *The Farmer's Boy* sold some twenty-six thousand copies in three years, making Bloomfield a celebrated peasant poet and the standard by which other working-class poets were measured.[10] One such poet writing at the same time was Thomas Bachelor (also spelled Batchelor), who is compared with the more popular Bloomfield in the 1805 *Eclectic Review.* Unlike Bloomfield, who left rural life, Bachelor was a tenant farmer under the Duke of Bedford at Lidlington, Bedfordshire, about forty miles north of

London.[11] Francis Russell, fifth Duke of Bedford, was a member of the Board of Agriculture, which set about surveying the nation's farms county by county in order to record normative and best practices and, through dissemination of knowledge, increase farm yields. Russell built and developed a model farm on his estate at Woburn. The annual Woburn sheep shearing, made famous by George Garrard's engraving, was a prominent regional fair for the promotion of new methods of breeding animals and improved modes of farming. The duke was also the first president of the Smithfield Club, founded at Smithfield market and designed to improve breeds for market. An examination of Bachelor's work takes us deep into the complex technological, economic, political, and moral forces at work in modernizing agriculture in the eighteenth and nineteenth centuries. As author of *General View of the Agriculture of the County of Bedford* for the Board of Agriculture, Bachelor was keenly aware of the role of the survey and the movement toward improvement. He generally agreed with the new agricultural methods; yet, as an examination of Bachelor's poem "The Progress of Agriculture, or Rural Survey" will make evident, peasants and animals alike were caught up in a modernization scheme that did not look out for their best interests.

Much of Bachelor's poem explains how the wild and rude land has been tamed: "around my natal soil I see / The bless'd effects of peaceful industry."[12] Patriotic "fair Freedom" gives a "generous hand," and "guards, improves, and dignifies the land" (75). The irregular ground is tamed: "There oft the plough has turn'd the glowing sand, / The spade, the pickaxe, smooth'd the rugged land" (77). Even the "industrious peasant" has a plot "richly supplying all domestic wants," and so the laborers give thanks "to gen'rous *Russel's* name" (78). As the poem extols the virtues of the Duke of Bedford and the Board of Agriculture's methods of improvement, it comes as little surprise that Russell's favorite livestock, the sheep, holds pride of place. After listing the treasures of other lands, such as India's spices, Gaul's wines, and East India's sugarcane, Bachelor praises the British Isles:

> Yet thy own wealth attracts the richest stores
> With power magnetic to thy favour'd shores.
> And chief thy flocks, that crown each mountain's brow,
> And deck each vale, from these thy riches flow;
> These meet my view, innumerous grazing wide,
> Their unshorn lambs yet sporting by their side.
> Some destin'd soon, by unrelenting fate,
> To smoke on tables of the rich and great;

But those of finest shape, and noblest size,
Again must view the vernal year arise,
Spread thy young progeny around the land,
And yield their fleeces to the shearer's hand. (83–84)

Interesting here is that a nation's wealth does not reside foremost in the labor of its people but in the value of its animals: "chief thy flocks." Like Bloomfield, Bachelor provides a pastoral line about the "sporting" spring lambs. And as in Bloomfield, the lambs are slaughtered, in this case "to smoke on tables of the rich and great." It is surprising that the poet who praises Russell and the Board of Agriculture clearly marks class differences. Animals and laborers are joined in two ways in this passage. First, both create material wealth that serves not themselves but the rich. Second, the sheep replace humans as the "richest stores" of the nation.

Despite his praise for agricultural improvements, Bachelor realizes that the laborer's work is not his own but is used to benefit the landed:

Still glow thy fields in summer's fruitful ray,
Thy harvests flourish, all thy meads are gay;
But not for me fair Nature spreads her store,
Life's smiling prospects must be mine no more!
. . . .
Monopoly has rear'd her gorgon head
To strike the source of rural comforts dead! (88–89)

The lambs give their life for the tables of the rich and the laborers give their toil to the same end. Bachelor's *General View of the Agriculture of the County of Bedford* provides an assessment of farms and labor that will ultimately solidify the power of those who hold a monopoly on the land. His poem, as a "rural survey," is yet another mode of counting the wealth of the landed and mapping the trajectory of prosperity from raw materials and labor to seats of power.

Greatest among the raw materials that build the wealth of the nation are its livestock, which become more valuable than the laborers. Bachelor is not simply using a rhetorical flourish or pastoral trope in calling sheep the "chief" treasure of the nation. From the mid-eighteenth century well into the nineteenth, tenant farmers, day laborers, and commoners were moved off arable land so that it could be used to pasture livestock, most commonly sheep. Despite the Duke of Bedford's investment in sheep, Bachelor laments the practice of clearing the land of its people and replacing them with animals:

Green pastures spread where harvests wont to smile,
Who change for herds, the life-supporting grain,
With *woolly tribes* displace the reaper train,
Who build a *palace* for the wealthier few,
But drive to squallid [*sic*] huts the *ruin'd* crew. (90)

Through a poetic syntactic chiasmus, Bachelor shows how sheep replace humans and grazing replaces harvesting. The "*woolly tribes*" of animals are put in place of the human tribes of laborers. Rather than drive sheep across the land, the landed drive to squalid huts the rural poor. The sheep are so highly valued that Bachelor depicts them as the means by which the British raised themselves from barbarism:

Yet sure yon patient, woolly tribes demand
The generous care of man's providing hand;
Unless for them his shivering limbs must bear
Th' enfeebling rigours of th' inclement year.
Or, wrapt in skins, the human form debase,
To a vile semblance of the savage race. (85)

In this poetic chiasmus, humans provide for the sheep, but it is the sheep that provide the wool that lifts humans from the debasement of wearing animal skins, like savages. In light of sheep replacing people on the landscape, the poet's statement that "the woolly tribes *demand* the generous care" of humans increasingly sounds like a demand made by the landed and imposed through the medium of the sheep upon the laborer. The Scottish Shetland, along with the rise of the Spanish Merino in Britain, increased the value of the British wool industry and made sheep more valuable and easier to manage than tenant farmers.

Of course, sheep in themselves are demanding, in that domesticated livestock require care. As the grazier and author William Youatt observed in a lecture delivered before the Society for the Diffusion of Useful Knowledge, later reprinted in his book on cattle:

> Cattle are like most other animals, the creatures of education and circumstances. We educate them to give us milk, and to acquire flesh and fat. . . . When he has lost the wild freedom of the forest, and become the slave of man, without acquiring the privilege of being his friend, or receiving instruction from him, instinct languishes, without being replaced by the semblance of reason. But when we press him into

our immediate service . . . he rapidly improves upon us; he is, in fact, altogether a different animal; when he receives a kind of culture at our hands, he seems to be enlightened with a ray of human reason, and warmed with a degree of human affection.[13]

Domesticated animals require care precisely because humans have domesticated them; they have "lost the wild freedom of the forest" in exchange for "a kind of culture at our hands." Humans have created the labor that the woolly tribe demands. The exact nature of these demands depends not simply on the animal but also on the human, whose selective breeding changes the nature of the animal and its needs. For example, when King George III began the rearing and experimental breeding of Spanish Merino sheep on his model farm at Windsor, he signaled a change in how the animal would be used.[14] He shifted slightly what sort of commodities would be gleaned from the animal. While many British breeds yield medium-grade wool, along with some respectable meat, the Merino (and to a lesser degree the Shetland) dominated the discourse on agriculture from the late eighteenth century onward. In summary, breeding practices promoted by the gentry fashioned the sort of demands the animals made on the laboring class, including replacing farmers with sheep even on arable lands, tending and feeding livestock (including diverting land for human food to land for fodder), and increasing the output of the British wool industry.

In Scotland, the replacement of humans with sheep was called "the clearances." As Bachelor mentions, such clearance occurred throughout Britain; however, its effects were particularly devastating in Scotland, which had seen changes in land management throughout the century after the Act of Union. In 1822 Major General David Stewart recorded the effects of the clearances on the land, animals, and people of Scotland in *Sketches of the Character, Manners, and Present State of the Highlanders of Scotland.* His descriptions arose from a request by his commanding officer in Scotland to provide a history of the people and terrain. Stewart supplemented his own recollections with those of other officers, including some who served in Scotland as early as the 1760s. Throughout the eighteenth century, Scotland was an experiment in military occupation that included building roads, bridges, and forts to bring the wayward Scots into the fold of the British nation.

Although the pastureland could be used for cultivation, farming required more capital and manpower than did the tending of livestock: "the pasture was more easily managed; that with ten men and twenty dogs, they would take care of all the sheep and cattle in the glen, which, under cultivation, supported 643 persons."[15] Stewart gives numerous examples of the working poor being run off tenant land. Occasionally they had enough time to pull up the boards of their cottages and move to the coast. Other times, their houses were burned and the families were left homeless and forced off the land. As with much of the English control over the Highlands, a new, "enlightened" system was imposed to increase productivity and to break the Scots of their independence and traditions. As Stewart recounts, "After the new system of managing lands and laying out farms had commenced in the Highlands, the ancient occupiers and cultivators were often overlooked by those who undertook to new-model gentlemen's estates. *Their* future happiness or misery formed no part of the new plans, and seemed as much disregarded as the fate of the ancient breed of horses and sheep. The old Highlanders were considered unfit for the new improvements."[16] Humans and animals that had developed alongside one another on the land had been replaced by "new improvements" in farming and breeding. The disregard for Scottish breeds was part of a larger rewriting of agriculture. Breeding of animals indicates a way of life on the land. The number of animals and their uses are determined by a way of thinking about agriculture that has unfolded over time. Usually, sheep grazed on land too poor or rocky or inaccessible for farming. This limited the number of animals. Likewise, the type of livestock (cattle, sheep, or hogs) and their size depended on the ecology of the landscape, the amount and types of food available for humans and animals, and the needs of people for milk or meat or wool. Just as Scottish animals were disregarded as part of an inefficient ancient system, so too the Highland farmer was considered unproductive. The humans and animals together materially signify a way of life. As the "enlightened" livestock edged out "the ancient breed," so too a way of life, a human-animal ecology, was transformed.

The Scottish politician and agricultural writer Sir John Sinclair played an important role in changing Highland agriculture. Sinclair's background proves representative of the shifts from an older mode of government and farming to an "improved" plan. He was the son of George Sinclair of the Earls of Caithness, in the northern mainland of Scotland. John Sinclair studied at the University of Edinburgh and then at Glasgow, where he was a pupil of Adam Smith. The management of national productivity evident in *The Wealth of Nations* (1776) is manifest in Sinclair's own work. Because of Caithness's rocky terrain, he focused on sheep production on his property, first trying five hundred Cheviot on his land and then increasing the number to six thousand. Then he set out to improve the wool of the native Shetland to be competitive with the Cheviot. In 1791 he established a sheep-shearing festival at Queensferry, and at the same time formed the Society for the Improvement of British Wool.[17] His material, political, and rhetorical investment in sheep helped further the clearances. As recorded in the *Biographical Dictionary of Eminent Scotsmen,* Sinclair claimed that "of all the means . . . of bringing a mountainous district to a profitable state, none is so peculiarly well calculated for that purpose as the rearing of a valuable breed of sheep. A small proportion alone . . . of such a description of country can be fit for grain; and in regard to cattle, for every pound of beef that can be produced in a hilly district, three pounds of mutton can be obtained, and there is the wool into the bargain."[18] Eventually, sheep occupied not only the hilly districts but the arable land and brought landowners greater prosperity and agricultural prominence than they would have obtained by leasing land to tenant farmers.

In accord with a number of agricultural treatises and surveys of the period, Sinclair created, administered, and managed the ambitious twenty-two-volume *Statistical Account of Scotland* (1791–99). His use of the word "statistical" was rather novel at the time and called for his explanation:

> Many people were at first surprised at my using the words "statistical" and "statistics," as it was supposed that some in our own language might have expressed the same meaning. But in the course of a very extensive tour through the northern parts of Europe, which I happened to take in 1786, I found that in Germany they were engaged in a species of political enquiry to which they had given the name "statistics," and though I apply a different meaning to that word—for by "statistical" is meant in Germany an inquiry for the purposes of ascertaining the political strength of a country or questions respecting matters of state—whereas the idea I annex to the term is an inquiry into the state of a country, for the purpose of ascertaining the quantum of happiness enjoyed by its inhabitants, and the means of its future improvement; but as I thought that a new word might attract more public attention, I resolved on adopting it, and I hope it is now completely naturalised and incorporated with our language.[19]

Sinclair relied primarily on roughly nine hundred local parish ministers to conduct the survey, with some help from hired agents. The survey included 160 questions covering geography, population, agriculture, and industry. While ostensibly seeking "the quantum of happiness enjoyed by its inhabitants," the survey actually recorded normative behavior of the population and laid the groundwork by which Sinclair and others would argue for improvements.

In 1793, while conducting the survey, Sinclair was instrumental in establishing the Board of Agriculture (officially chartered as the Board or Society for the Encouragement of Agriculture and Internal Improvement), and he served as its first president. With a number of individuals and local societies advocating improvement of agriculture and livestock, it was simply a matter of time before matters were centralized and nationalized in such a board. According to the biography of Sinclair by his son, Rev. John Sinclair, the idea of forming a national board first occurred in Sinclair's essay on the corn bill, "An Address to the Landed Interest on the Corn Bill" (1791). Sinclair, his son tells us, had "been deeply impressed by the fact, that while we were alleged to be dependent on foreign countries for food, there existed in England alone twelve millions of acres almost in a state of nature; and that many statesmen were looking helplessly for subsistence to other countries, while they overlooked the abundant capabilities of their own." According to the biography, Prime Minister Pitt was impressed with this information, which proved "important to the safety and independence of the kingdom."[20]

From the Society for the Improvement of British Wool, to the *Statistical Account,* to the Board of Agriculture, Sinclair was fundamentally interested in what Michel Foucault would call biopolitics and biopower. The basic parameters of these terms are outlined in Foucault's lecture of 17 March 1976, published in *Society Must Be Defended: Lectures at the Collège de France, 1975–1976.* Biopower is the event of the biological tended by state control. The political, rational, legal, and rhetorical justification for such control Foucault calls biopolitics: "Biopolitics deals with the population, with the population as political problem, as a problem that is at once scientific and political, as a biological problem and as power's problem." While regulating the citizens as a mass has been a general concern of government throughout the ages, Foucault claims that the eighteenth century saw a rise in technologies used for this purpose. Such regulation was not the age-old negative modality of regulating death: the right of a state to take the life of a citizen who violates its laws and to send citizens to war. New state powers included "mechanisms with a certain number of functions that are very different from the functions of disciplinary mechanisms. The mechanisms introduced by biopolitics include forecasts, statistical estimates, and overall measures. . . . [They] intervene at the level at which these general phenomena are determined, to intervene at the level of their generality." At the turn of the nineteenth century a number of mechanical, information, and economic technologies aligned to "maintain an average, establish a sort of homeostasis" for the population.[21] The state co-opts and fosters these technologies to the end of greater control of the population. We see the concern for measuring, forecasting, and regulating population played out in the debate between Godwin and Malthus, then in Southey's 1803 review of Malthus in the *Annual Review* (published in 1804). The issue is central to Goldsmith's *Deserted Village* and George Crabbe's *The Village* and *Tales in Verse.* By claiming that Sinclair's labors can be understood through the modality of biopower, I am also making a larger claim about the link between humans and animals.

Sinclair's concern for yields of wool and meat, his development of statistical reports and agricultural habits, and his development of a board that promoted norms and improvements in agriculture, including livestock, all link the health of British citizens with the cultivation of British animals. Foucault considers the regulation of sexuality a pivotal example of biopower: "It is, I think, the privileged position it [sexuality] occupies between organism and population, between the body and general phenomena, that explains the extreme emphasis placed upon sexuality in the nineteenth century."[22] I would simply add that food, too, functions as a mode of regulation between "the body and general phenomena" of the masses. This is not "simply" the problem of feeding a people and preventing famine. It is also an issue of what sort of food, its cost and distribution.[23] While Sinclair and his fellow agriculturalists, such as Arthur Young and Francis Russell, considered variables of agriculture and their effects in a general sense, the working-class laborer felt the material consequences of such regulation. It is common for agricultural improvers to deride farmers for their backward practices and for resisting innovation. This resistance can more properly be understood as a problem of vulnerability. If crops fail or livestock do not fatten on a model farm, the wealthy landowner loses little. In contrast, the tenant farmer prefers to trust a tradition that has been tried on his particular locale and soil in varying weather conditions over decades rather than gamble his crops and livestock on innovation. Agricultural innovators can think in terms of "forecasts, statistical estimates, and overall measures," while tenant farmers—particularly the less wealthy sort—and day laborers are tied to their particular patch of earth. These laborers are more concerned with immanent vulnerability than with the abstract value of agriculture as promoting the "safety and independence of the kingdom."[24]

It is not that gentlemen agriculturalists and innovators are unaware of the specificity of their crops and livestock. Indeed, innovators like Bakewell and Townsend pay close attention to the details of their sheep and turnips, respectively. However, they understand and fashion these details through the technologies of biopower. In contrast, the laboring classes have a visceral connection to the animals through their bodily labor and daily contact. Their connection to the animal is through a physicality and vulnerability. Anxieties over animal death manifest in the picturesque and in select working-class poetry reflect this mutual vulnerability between workers and livestock. As Wolfe puts it, workers share "the physical exposure to vulnerability and mortality that we suffer because we, like animals, are embodied beings."[25] The argument of a shared animality is based not on the animal's capabilities or on that of humans but on a mutual fragility. It is this same lack of power that provoked Jeremy Bentham's question—"Can they suffer?"—that establishes a moral relationship between humans and animals based not on ability or productivity but, as Jacques Derrida observes, on an incapacity to act.[26]

The capacity to suffer is a potent element of Robert Burns's poetry. Much can be made of how the poet links the fate of the Scottish people to the death and resurrection of rye in "John Barleycorn." Plowed down, cudgeled, crushed by stones, Barleycorn's blood becomes the liquid that pours through the Scotsman until "a man forget his woe," and John Barleycorn's "great posterity / Ne'er fail in old Scotland!"[27] Burns connects the people not only to the crops but also to a variety of animals. Best known, perhaps, is "To a Mouse, On Turning Up Her Nest with the Plough" (1785). As winter comes and the fields are barren, the mouse finds a home in the stubble of the field. The narrator's plow upturns the nest, "proving foresight may be vain; / The best-laid schemes o' mice an' men / Gang aft agley." The speaker addresses the mouse and explains that if it feels pain, consider the worse fate of the laborer:

> Still thou art blest, compar'd wi' me!
> The present only toucheth thee:
> But, Och! I backward cast my e'e.
>   On prospects drear!
> An' forward, tho' I canna see,
>   I guess an' fear![28]

The forward and backward glance of the laborer reflects the uncertainty of his fate and the fate of his crops. The temporal glance is doubled by a physical look backward and forward as he aligns his plow to create evenly spaced and equally deep furrows in the soil. The physical act of his plowing coincides with his anxiety over future crops and his ability to feed himself and his family.[29] The mouse and human are aligned in their vulnerability, but because of his capacity for forethought, the human feels the anxiety more intensely.

The odd singularity of details in Burns's encounter with a field mouse reveals a particular intensity and sympathy for the lives of the working poor. From lacunae in representing death to anxious moments of maintaining a livelihood, representations of farm animals (or, in the case of Burns's mouse, animals on the farm) reveal a larger set of forces at work on beasts and citizens alike. Laborers were often incapable of directly challenging the construction of biopower in the rise and reach of British nationalism; however, they could deflect or adversely reflect these powers by representing the state of affairs for themselves and the animals that were their responsibility and their livelihood.

Chapter FIVE

The Rights of Man and the Rights of Animality at the End of the Eighteenth Century

PIERRE SERNA
Translated by VITO CAIATI AND JOAN B. LANDES

Since antiquity, philosophers have acknowledged Aristotle's claim that man is a political animal.[1] Yet the converse claim that animals are political subjects remained controversial at the end of the eighteenth century, when the French Revolution shook the foundations of French society by according each individual an equal place within a radically transformed social, political, and cultural order. Abolishing the old order of inherited privilege, the revolutionaries transformed the basis of society from biological markers of birth to the legal equality of all citizens. The ensuing debates over the status of citizenship provoked a series of intense reflections regarding the status of animals. Implicitly, the idea of human rights raised the question of whether the concept of rights should affect the world of animals, understood according to their physiology, status, and essence. What follows charts how the rights of animality—that is, rights shared by nonhuman as well as human animals—were addressed by the new citizens of revolutionary France in the newly declared French Republic, particularly those belonging to the new community of "republican" science.

It is noteworthy that the noun "animality" is still used colloquially to refer to animal nature or character, suggesting a prehuman, physical, or instinctive behavior. However, the term derives from the Latin *animalis,* meaning "animate," or "living." In certain late eighteenth- and early nineteenth-century European arguments about animality, one discovers the articulation as well as the refusal of a broader conception of rights shared by both humans and animals. In brief, the denial of rights came to be tied to an emergent racial ideology implicating both animals and humans.[2]

In 1792, the creation of the First French Republic mobilized republican scientists on behalf of a "nation in danger." Propelled by a series of foreign and civil wars lasting until the final defeat of Napoleon in 1815, an entire generation contributed to the construction of an ethic that placed knowledge in the service of the nation and the developing republic.[3] Among its many projects, the new republican legislature (known as the Convention) founded the Museum of Natural History, intended to be a place of conservation, care, and study of natural organisms. This crucial venture was tied to the infant republic's commitment to natural rights through the goal of protecting life-forms and placing them on display for the benefit of all its citizens. In addition, the museum sponsored a series of public lectures on the mysteries of nature and man's place in the world—a world for which man, its most advanced creature, was deemed responsible. Located at the center of broad-ranging intellectual speculation, and supported by an important publishing program, the animals were the subjects of special attention in the promotion of knowledge about man and his development.[4]

Notwithstanding considerable hesitation and controversy generated by this exploration, republican scientists sought to establish a rapport between the world of animals and that of man. Historians of science have located and explained these debates within a paradigm of emergent scientific knowledge and research.[5] Other historians have addressed the history of scientific controversies tied to the animal question within a social history of scientific institutions, showing how a narrow focus on epistemological questions obscures the ambitions and strategies of distinctive groups.[6] Moreover, a history of the networks of the scientific world has appropriately linked the issues of intellectual dueling between different camps of scientists with corresponding battles over political positioning. The translation of vitalist or organicist theories into ideas of republican organization during this age, for example, reveals how Napoleon Bonaparte and his supporters aimed to reintroduce social stratification within the purportedly egalitarian republic by anchoring the former in the laws of nature. According to the "irrefutable" logic of science, Bonapartists asserted, only the strong command.[7] Yet, until now, historians of politics during the Revolution have expressed little interest in the place of beasts—particularly, in the latter's specificity of "being animal"—concentrating instead on more traditional subjects, such as the exercise of citizenship in all its aspects.[8] In contrast, this chapter places animals at the center of inquiry that occurred during the Consulate, when considerable attempts were made to stabilize and tame revolutionary principles.[9] During this period there was an increasing emphasis in scientific paradigms on classification and utilitarian order, which was accompanied by an insistence on the difference rather than the philosophical and metaphysical similarities between man, animal, and nature.[10]

This chapter focuses on texts by two authors working during the Consulate. Jean-Claude de Lamétherie's writings located the question of the relationship between animality and humanity at the center of an increasingly conservative social organization. Alternatively, Jean-Baptiste Salaville cast a disturbing light on the relationship between man and animal. Seemingly more "animal-phobic" at first glance, Salaville was in fact more committed to the revolutionary ideals of liberty than to the ascendant modes of classification and of natural and hierarchical stigmatization. Both authors were concerned with issues of citizenship and social order. They raised essential questions about the ontological distinctions between human and animal, and in turn crudely posed the issue of subordinate or secondary citizenship for a group whose status was being hotly debated during these same years: (black) slaves. By focusing on animals, the authors put forward questions about the rights of all beings and revealed the intellectual justifications offered for violence done to inferiors through a definition of otherness. By examining how distinctions were drawn between bestiality and humanity, we discover some of the paradoxical ways in which the goals of social reproduction worked to undermine the democratization of society.

Establishing a Human Farm in the Early Years of the French Republic

The manuscripts of the Museum of Natural History include a letter signed by the French author and botanist Jacques-Henri Bernardin de Saint Pierre, dated 21 January 1793, in which Bernardin attempted to save the life of the "rhinoceros and its companions" of the king's menagerie at Versailles, pleading their cause and defending the idea of establishing "a menagerie at the national garden."[11] The Museum of Natural History was founded on 10 June 1793, with zoology occupying an important place in its courses of study. The chair of invertebrate animals was given to Jean-Baptiste Lamarck, and that of the vertebrates to Étienne Geoffroy Saint-Hilaire, who would eventually take charge of the new public menagerie established in 1794.[12] However, the menagerie's situation remained uncertain until the decree by the revolutionary government, issued on 27 Floréal, Year II (16 May 1794),[13] officially authorizing it by ordering the construction of some enclosures. Toward the end of the year, the menagerie contained about sixty-five mammals and twenty-five birds.[14]

The political deputy Antoine-Clair Thibaudeau's report of 21 Frimaire, Year III (11 December 1794), on behalf of the Committee of Public Instruction and the Committee of Finance, tied the menagerie to instruction: it was "constructed so that animals of any species may enjoy all the freedom that is consistent with the safety of spectators, so that we may study their customs, their habits, and their intelligence."[15] In December 1795 the naturalist Bernard Germain Étienne de Lacépède[16] wrote a letter on "the public establishments . . . known as the menagerie," in which he asked that exotic animals be sought less for their unusual appearance than for their ability to be "acclimated," and advised that they "may be in whatever way free in larger or smaller enclosures, so that the images of coercion or the appearance of slavery may be as far as possible from the eyes of a free people."[17] The menagerie was not completed until 1798, following the confiscation of the collection of the Stadtholder William V of Holland by the invading French revolutionary armies and the arrival of a mated pair of elephants from the Stadtholder's former menagerie at Loo.

Thereafter, despite the vicissitudes stemming from the endemic crisis of the Directory that required the sale of some animals and the culling of the less valuable, the menagerie grew and became an important place for the dissemination of knowledge. In Year VIII (1799/1800), the museum's first librarian, Georges Toscan, listed a hundred residents in the menagerie.[18] The dramatic political changes wrought by the coup of 18 Brumaire (9–10 November 1799), overthrowing the Directory and establishing the Consulate with Napoleon Bonaparte as first consul, proved to be beneficial to the stewardship of the new institution.[19] The ensuing intellectual discussions surrounding animals served the professors at the museum, who were awarded choice posts within the new order of knowledge. Lacépède, chairman of the French Senate and chair of reptiles and fishes at the museum, protected them and defended their publications.[20] Tremendous interest in all the natural science disciplines stimulated wide discussions on such topics as the transformation of human intelligence, spontaneous generation, the extinction of species, the emergence of new forms of life, the relationship between zoological taxonomy and chemical theories and mineralogy, and especially the history of the earth as related to hydrology and geology.

Alongside this general interest in vital phenomena, and in the animal world in particular, the 1790s witnessed a scientific infatuation with apes. Since the 1750s there had been continuous interest in great apes, but the rise of anthropology, the return of several important maritime expeditions with live specimens, and many polemical publications on the origin of the orangutan and its strange resemblance to man provoked new discussions about the identity of man, his origins, and his organization.[21] The luxurious work of the artist and naturalist

Jean-Baptiste Audebert, *Histoire naturelle des singes et des makis* (Natural history of monkeys and lemurs) (1800), based on the collections of the museum, received the intellectual backing of Lamarck and Lacépède. In perfectly mastered rhetoric, Audebert's preliminary discourse awakened readers to the proximity of man and ape but recalled at the same time all the trouble implicit in proclaiming the similarity of the two beings: "Man, astonished at the unexpected appearance of these animals, and so to speak, ashamed to acknowledge the very great number of his links to a being that presents the features of degraded humanity, may grant to apes a nature and understanding superior to those of other animals. Hence these wonderful stories, these daydreams, fruits of a deceitful imagination accumulated by his predecessors in speaking of these animals. The Orangutan, in particular, has long been considered a half-human species."[22]

Audebert clearly did not believe, but all the same related with delight, popular stories about the abduction of "black women" by apes, observing that a variety of the human species, the Chacrelas,[23] had been mistaken for this animal. As an enemy of prejudice, Audebert could not admit the existence of a "species midway between man and animals." On the other hand, he explained his approach to determining the difference between types of beings through the calculation of the facial angle: "passing from the Europeans to lesser-known people, such as the Kalmouks [a Russian Mongol-speaking people] and the blacks of Guinea, this line gradually moves away from the perpendicular. And if one descends from man to bird, whose facial line is nearly horizontal, one will see that in considering the animals through this point of view, the Orangutan, occupies the first rank after the man."[24]

Thus Audebert suggested a classification based on human exclusivity as well as a ranking of different groups of humans. Inscribed within a profoundly revitalized and still open field of inquiry, his writings nonetheless reflected a growing rigidity in intellectual positions. This rigidity culminated in an explicitly racialist vision of the different human species, posited by other authors during the early years of the Consulate, when slavery and colonialism were reestablished, and continuing more emphatically in subsequent decades of the nineteenth century.[25]

Lamétherie, or The Old Ape Conservative

In 1802 Jean-Claude de Lamétherie published *De l'homme considéré moralement, de ses mœurs et de celles des animaux* (Of man considered morally, his customs and those of animals). The editor from 1785, and then co-owner, of the *Journal de Physique* (Journal of physics),[26] and a member of the liberal Club of '89, Lamétherie had initially endorsed the constitutional monarchy established at the beginning of the Revolution. However, he became more critical of the direction of the Revolution, ultimately fleeing the capital during the Terror and resuming publication of the *Journal de Physique* only in 1797. A pioneer in soil science, Lamétherie was a specialist in the history of the Paris basin and had a decisive influence on Lamarck's transformationist views. He took an early interest in the history of the earth, as well as in the study of the nature of man and his place in the animal kingdom. His materialist orientation is evident even in his works dating from 1780, where he compared animals and plants to superb hydraulic machines, driven by different motor forces "arranged in complex mechanisms."[27] Lamétherie's *De l'homme* applied the knowledge gleaned from more than two decades of intense reflection and research in natural history to the field of political thought. But *De l'homme* is a curious mixture of learned observations and prejudice, reflecting the author's conservative politics.

Lamétherie summarized his argument in a long preface. He maintained that two categories of beings exist: men in society and animals, including men in nature. Anxiety, boredom, and the constant search for pleasure characterize the former, while play alone occupies the latter. Animals and men in nature live in the contentment of the present, while men coming together in society have developed the perfectibility that constitutes the surest sign of humanity, as well as its corollaries, anxiety and the unquenchable thirst for possession.[28] For Lamétherie, the proximity between human beings and animals arose from the similar mechanical organization of their bodies, "a true machine whose moving principle is found in the nervous system," driven by the movement of a subtle fluid whose nature remained unknown to scientists. This claim led to another, more explicit one: "Man belongs to the ape family, was originally covered with hair like them, lived in hot countries, and ate only fruit, seeds, sometimes herbs." Thus "man, who is the leading species of ape, must be placed at the apex of the scale of animals."[29] Through a combination of historical circumstances, man became the absolute master of the earth's surface, while retaining his animal nature. Lamétherie distinguished six stages of evolution, from the meeting of male and female up to the union of men in civilized societies who build cities and communicate through the arts and sciences—the last being beyond the reach of the apes but related to them by this chain of historical development. Like Audebert, Lamétherie used the different facial angles of humans and animals as an index of human perfection and a measure of the hierarchy of living beings.[30]

At different points in the text, Lamétherie stressed the similarity between natural man and animals, and more specifically apes, as when speaking of their desires, their survival through the organization of a territory, and their common courage in situations of conflict. They all shared a natural liberty that man renounced in order to associate himself with others of his kind. Lamétherie said of this shared identity, "All animals outwardly reveal their feelings; man does not hide his more when he is not restrained by social considerations."[31]

How is the reader to understand this transformation of man, who is so changed that he has forgotten and sometimes vigorously denies his animal being? There existed an indefinite period, where for reasons of survival men abandoned their frugivorous status to become, against their nature, carnivores. That epoch transformed the habits of humans, who from then on ruthlessly slaughtered other animals. Cold-blooded violence and war caused this alteration. While conferring on them more and more intelligence, this passage also plunged humans into a state of permanent aggression.[32] According to Lamétherie, the human being assumed the character of carnivorous animals; with overheated desires, he joined with others to make war. To be sure, his level of technological sophistication became more perfect. But Lamétherie pessimistically linked the level of civilization with the violence among nations, describing the phenomenon of killing in battle as being agitated by the power of feeling and occurring under "the influence of a greater pleasure, that of defending the nation."[33] As the inheritor of the eighteenth century's commitment to enlightenment, Lamétherie still hoped, all the same, that education would accomplish its work, and that the "frugivorous" roots of man would not be totally lost. He described several acts of kindness and philanthropy, along with a few heroes in history, which traced a possible, more hopeful path for humanity.

Principes de la philosophie naturelle dans lesquels on cherche à déterminer les degrés de certitude et de probabilité des connaissances humaines, a text published anonymously in 1787 in Geneva but attributed to Lamétherie, offers insights into the coherence of the author's radical materialist thought. This work pleads stirringly for a total reconsideration of the organization of living beings and of the human-animal relation. Man is "an animal like any other," simply enjoying a more refined organization thanks to his much greater perfectibility. He must understand that every animal endowed with sensitivity is his equal and, in principle, enjoys the same natural rights as human beings. These rights are innovatively divided into two categories: "One which concerns all the animals, which we will call the right of animality, and the other, which only refers to each species of animal: the latter takes the name of humanity, when it is a question of man." Even more boldly, this text developed the idea of a necessary love, the kindly esteem that man must convey to each animal; and it specified under what conditions it is better to rescue an animal than a human being, following a law of equity among living beings. Logically, it criticized those who had so forgotten their duties that they enslaved animals by working them too hard, killing them for consumption, or hunting or torturing them for pleasure, as for example in Spain during bullfights. In this argument, as in Lamétherie's later writings, it is the carnivorous animals that cause problems. Has man imitated them? By what measures did he become the master of the earth? By whose authority did he reduce domestic animals to slavery, driving wild animals as far away as possible, and creating the conditions of "a continual state of war with other animals"?[34] In *Principes,* Lamétherie went even further, contending that man had no moral right to kill an animal or to eat it. Men were once and must again become frugivorous. They have no proclivity to remain carnivorous, and it suffices to note that plants are sufficient for their health, so that they should stop using their teeth like the carnivorous animals.[35] Similarly, work relations should be regulated according to a strict law of reciprocity:

> The animal has no reason to complain, but one should take his strength into consideration and not impose on him work, which he is too weak to perform. Man should treat him as another self. He may not overcharge him with excessive tasks, and respecting his happiness [*jouissances*], he will allot some time to his rest and pleasure [*plaisir*], as he does for himself. He may also make use of the dairy products of some and eat the eggs of others, since in doing so, he does harm neither to their joy nor their existence. So it would not injure a ewe to shear her fleece nor any less a silkworm to take its cocoon. Finally he will extend his care to the animals until they die naturally. At no moment is he permitted to end their lives, to eat their flesh.[36]

Notably, Lamétherie not only addressed a projected revolution in the relations between man and animals but also traced in a resolutely modern manner the new conditions of an ecological and political order. Even as European man extended his grip on the land, Lamétherie fretted, he would not stop usurping territories until none remained. It was necessary to stop colonial adventures. Otherwise, "the other species of animals will be sacrificed more and more to man, and a great number will disappear from the face of the earth."[37] There would thus come a time when men became so numerous that they would lack basic needs, and another time when America, already liberated by its revolution, would be followed by Asia and India, where the harbingers of revolution were already visible. Accordingly, it was essential to anticipate these impending environmental and political dangers, whose consequences consisted of an overpopulation that would inevitably destroy the relationship between population and subsistence. Thus, just as he does with domestic animals, man must control his own reproduction in order to maintain a balance between nature and human society.[38] In conclusion, the social order required that children be raised as vegetarians, without being exposed to the killing of animals. It required that man respect the right of animality and respect animals' right to enjoy, as he does, whatever nature produces. The social order would be respected, reform made possible, and revolution avoided, Lamétherie maintained in 1787, only if these conditions were met.

Fifteen years later, in 1802, in the aftermath of the revolutionary upheaval he had earlier feared, the ecologist Lamétherie remained steadfast in his materialist approach toward man, but his perspective on social organization had evolved considerably toward a more conservative position. However, his arguments are a reminder that materialism is not necessarily associated with radical "leftist" politics. On the one hand, Lamétherie's scientific materialism drove him to demand a treatment of animals that was at the same time not demeaning to man, making man a sort of superior ape. This position indicated a shifting of his observations on man's immediate inferiors within a new system of relations that included social and biological dimensions. On the other hand, as it matured, Lamétherie's conservatism was in many ways similar to attitudes associated with the emerging doctrine of "social racism," fitting well the apprehensive mood of Consulate discourses concerning the order of men in society and Napoleon's restoration of French participation in the slave trade and colonial rule, a product of the new regime's efforts to consolidate its authority against democratic participation.

Indeed, beginning with Napoleon's coup d'état, political discourse hardened markedly against the popular classes, who were to be placed under ever-greater surveillance. With respect to social representations, a waterproof border was constructed between different classes, such that popular groups appeared as foreigners in society (*la cité*), implacably defined by their social inferiority. Men and women of the people were treated like another race, beings of a different "nature." This mode of discourse, which germinated in the conservative discourse of Thermidor and the establishment of the Directory (following the toppling of Robespierre and other leading members of the most radical phase of the Revolution), acquired an additional form of justification beginning in 1802 with Napoleon's reestablishment of slavery. By the same logic of exclusion, the common people lived in the towns (*villes*), while black slaves lived in the islands. Social exclusion disturbingly mirrored proslavery racism, appearing as what was being called "social racism" in order to indicate the attempt to naturalize the inferiority of the popular classes according to a pseudoscientific classification.[39]

It is here that a history of the scientific and moral debate on animals takes a fascinating turn, because it reveals a sort of double language within republican science, caught between the idea of polygenism and monogenism, taken in their political and social consequences. In the first case, science would authorize a progressive classification inside the lifeworld at the risk of inventing an indisputable "because demonstrated" inferiority of certain human groups. In the second case, continuing Buffon's legacy of monogenism, there would be a desire to refuse a classification between human beings according to their irreducible difference from animals.[40] The intellectual world vacillated between these perspectives, which in turn induced legitimating postures for legislators, resulting in either an enlargement or a shrinking of republican circles. In this light, the great debate among French scientists during the Directory amounted to thinking about the nature of living beings by recasting the categories of distinction and the frontiers between the animal and human worlds. Such a recasting could lead either to some form of all-encompassing legitimation of the universality of humankind within a circle of participatory citizenship (with the notable exclusion of women, young people, and people of color) or, with equal reason, to normalizing what we might call "infra-citizenship, which Elizabeth Cohen discusses under the concept of "semi-citizen[ship]." Under these circumstances, we can suddenly comprehend the major significance of scientific speech and its vulgarization in the debate over the civic and political recasting of citizenship, all of which played a role in the redefinition of what it meant to be French between 1795, with the beginning of the Directory, and the creation of the empire in 1804.[41]

The Napoleonic regime's evolving conservatism, including its restoration of slavery and its decision to entrust power for life to the ruler, ended any hopes for social and political democratization. These developments also affected Parisian scientific circles. Under the sway of political considerations, scientists subsumed the inquiry of animal nature into the regime's programs to build a rigid social and police order. Ultimately, Lamétherie carried out a spatial-social-temporal rapprochement between the undifferentiated order of apes and natural man in relation to civilized man, and, within the category of civilized man, between the working and dominant classes. Convinced that traits were transmitted by heredity from generation to generation, creating a hierarchical classification among men, Lamétherie implicitly rejected the idea of human equality, mentioning on several occasions distances between certain individuals so great that one could believe they belonged to different species: for example, between "a Papuan and Homer," or "a disgusting Hottentot and a Laïs."[42] In fact, his considerations about the mode of human functioning in society, as he admitted further on in the book, pertained only to rich men whose wealth provided abundantly what they desired: "It is he with whom we concern ourselves principally in this work, while the man attached daily to the soil or to his workshop . . . is condemned to work. . . . If his labor leaves him the time to cultivate his mind a little, he may acquire a reduced acquaintance of the great system of existing beings."[43]

These arguments reveal Lamétherie's attitude toward the supposed lesser sensitivity, kindness, and foresight of the lower classes. "We can apply to the barbarous and little civilized people what we say about the lower classes of civilized peoples," that they are a kind of scarcely polished social animal. Similarly, while praising foresight and a concern for the future as the hallmark of a higher humanity, Lamétherie stigmatized the humble classes and compared the poor to cruel animals. "The lower classes are more ferocious than the others; they willingly amuse themselves by making an animal suffer. What we call education among the upper classes softens their character a little."[44]

As in his general classification of apes, Lamétherie argued that superiors and inferiors formed two classes. In the "social state," he wrote, "there are also two absolutely distinct classes of men. One is made up of manual workers whose bodies are degraded by toil and who are incapable of morally elevating themselves; the others possess enough money to enlarge their minds, perfect their hearts, and strengthen their bodies, as in the world of animals."[45] Consistent with Napoleon's oft-quoted phrase on careers open to the talented, Lamétherie conceded that each individual could rise by his talents, acknowledged that the rich were not immune to defects, and agreed that common sense was acceptable when rid of the "meanness of the inferior classes." He therefore translated his materialist observations on man's animal essence into desired social and political arrangements. These views clearly expressed his fear of disadvantaged groups, which he explicitly identified as brutes with animal instincts who required monitoring and training within the republic.[46]

Jean-Baptiste Salaville, Republican Defender of Man Against the Animals

The most perceptive critic of the man of science and conservative zeal was the republican author Jean-Baptiste Salaville.[47] In the strange 1805 text *De l'homme et des animaux* (Of man and animals), which is partly a competitive academic essay, partly a systematic refutation of Lamétherie's theses, Salaville reflects on a question originally posed by the National Institute, the revolutionary successor of the Royal Academy of Sciences. Each year, one of the three classes or subdivisions of the National Institute organized a competition, which took the form of a question officially posed in public meetings and was richly endowed with significant sums. Thus, on 17 Messidor, Year X (6 July 1802), the class of moral and political science proposed as a subject for a prize in morals and politics the following question: "At what point does the cruel [*barbares*] treatment of animals become a matter of interest for public morals, and would it be advisable to make laws in this regard?"[48]

Salaville's essay contrasted the English parliament's mature debate on the subject of animal welfare with what he regarded as the primitive cruelty of Spanish bullfights.[49] Salaville clarified two opposing positions that he considered erroneous and harmful to correct thinking about the status of man. The human being, he argued, is neither the result of the religious dichotomy strictly separating the soul from the body, creating the certainty of a harmful flaw, nor composed of matter that would make him "an animal wiser than others, but for the rest not different at all." Salaville firmly criticized aspects of these viewpoints in Lamétherie's work. Yet Salaville was a man of letters, not a scientist, and was therefore unable to put forward another definition of the human with the same weight as that proposed by Lamétherie, the established scientist and multidisciplinary scholar. He appealed to the reader's trust that the moral dignity of man was sufficient basis for irrevocably distinguishing man from the apes.

Before analyzing Salaville's essay, it is necessary to recall briefly the somewhat chaotic sequence of events concerning this competition. After proposing the subject on 6 July 1802, the class of moral and political sciences, the well-known site of republican resistance to the first consul's authoritarianism, was dissolved by a decree of 3 Pluviose, Year XI (23 January 1803). The members of the sixth section, called "morals," had come up with the idea for this subject. The members of this section were Jacques-Henri Bernardin de Saint Pierre, mentioned above as the fierce protector of the king's rhinoceros and other animals in 1793, Louis-Sébastien Mercier, well known for his aversion to violence against animals, the abbé Grégoire, appreciated by philanthropists for his steadfast attention to the weakest members of society,[50] Louis-Marie de La Révellière-Lépeaux, architect of the natural religion "theophilanthrophy," Joseph Lakanal, and Jacques-André Naigeon. Despite the closing of the class, twenty-six competitors sent their dissertations (*mémoires*), assigned in the order of their arrival by the secretary of the class. Number 14 has been lost by the commissioners of the contest, and a final submission was not entered. What would become of these dissertations, since the "class morale" no longer existed? The former class of history and ancient literature received these texts instead.[51] Each was read, and the members of the jury—Jean Philippe Garran, Pierre Samuel Dupont de Nemours, François Emmanuel de Toulongeon, and Samuel Ustazade Silvestre de Sacy—signed each submission. Finally, during the public session of 2 Germinal,

Year XII (September 1803), no first prize was issued, but honorable mention was awarded to four of the submissions.[52] Unlike others, this competition was not decided, and no laureate was awarded the gold medal, on the grounds that the jury was persuaded that the best dissertations would not be able to clarify the subject in a more relevant way. This was a rather strange position on a subject so directly at the heart of investigations by republican scientists. In contrast, less polemical subjects, like the nature of fattening animals that hibernate, were reoffered. Was this the result of a deep conviction on the part of the jury? Or was it an act of self-censorship by the jury, since the political climate of 1804 was no longer sympathetic to arguments about the rights of the weakest members of society?

Salaville's essay was thus not favorably received. He published it nevertheless in 1805.[53] In the introduction of the printed text, he hypothesized about the decision not to award the first prize. Was it because the jury had decided that from now on it preferred a secondary question, or had it deliberately refused to recognize Salaville's talent and ideas, which went against the well-known culture of the jury's commissioners, all fervent admirers of the philosophy of the Enlightenment, as represented by Étienne Bonnot de Condillac, whom Salaville had contradicted point by point?[54] Beyond the "normal" paranoia of any writer facing the exacerbated conditions of a competition occuring in the republic of letters at the end of the eighteenth century, it is impossible to know definitively why the question posed in this competition was not reintroduced. Of typical relevance to the concerns of the members of the former class of moral and political science, it no longer had a place in the new order. To pose a question that could logically have prompted a reflection on the servitude of men (who would be able to be used like beasts after the reestablishment of slavery) was to risk charges of subversion in a country that was being dragged, in early 1802, into the freedom-killing police state of Bonaparte.

Salaville's text is one reflection of these circumstances. If it doesn't shine for its scientific mastery, it poses politically pertinent and subtly dissident questions only a few months after Napoleon's coronation as emperor. In contrast to Lamétherie's argument, Salaville's reflections on rights and all their consequences are considerably more pertinent to the implications of questions concerning the right of animals. The first element of his proof is based on the inconsistency of most proponents of the thesis about the proximity between man and animals. If this thesis were accepted, nothing would justify the eating of animals. Accordingly, one would have to conclude that the efficacy of a continuing crime, a kind of continuous murder of animals, signified a mandatory tragic fate for man, who is constrained to become an assassin in order to live.[55] Either the kinship is real, and one must immediately stop feeding on animals, or animals are of a different nature, and man is right to consume them. The argument, hitherto paradoxical, appears a little better supported when one begins to tackle the radical difference between animals and men by way of their rights.

Salaville raised the question of positive law early on. Since positive law results from an agreement freely and clearly entered into by two parties, it is evident that no such accord will ever be concluded between humans and animals. And because animals are unable to claim such a right, man cannot be considered guilty of any violation of a right they do not have.[56] Salaville here referred specifically to a kind of sensitivity (*sensibilité*) or form of animal intelligence such as that proposed in the new edition of Charles-Georges Le Roy's *Lettres sur les animaux* (Letters on animals), published in 1802 by Pierre Roux-Fazillac, supporting the idea of the perfectibility of animals on the basis of long observation of their behavior while being tracked by hunters and their ability to adapt when confronted with threats. Salaville asked again whether animals have natural rights. He noted the natural inability of animals to apply such rights: "animals do not have any right to have rights, since they can neither provide nor establish any of these; thus, their nature lacks that which may command respect for them."[57]

This is a crucial yet ambiguous point in Salaville's argument. For him, it is understood that animals do not have natural rights and thus cannot become subjects of positive law, not knowing or being able to express these original and fundamental rights. This does not go without posing a problem of interpretation. The author leaves us to suppose that it is the explicit absence of relations (at least those constituted by speech) between living beings of different natures that prevents animals from gaining entry into the circle of legal protection. Thus natural rights cannot be founded on an essence attached since birth to every living being. Rather, natural rights are the expression of a purely cultural act, a shared language, as they already mark entrance into the circle of human relations. As the expression of a human faculty constructed by education, language is not a birthright. This question, which Salaville left unanswered, becomes all the more disconcerting in the case of a child or a mentally handicapped person who lacks the capacity to speak. Does such a person possess natural rights, which makes him the subject of law, even though, in the first instance, he doesn't know how to speak, and, in the second, he is reduced to a vegetative life? Is he first an animal by nature? Or is it rather because of his potential to become a person, in the case of the newborn, or by virtue of his birth as a human being, which confirms on him by genealogical obviousness the title of human, that the question of such an individual's rights—natural and humanly natural—is not even posed? These are the questions that Salaville does not raise but that the reader can legitimately pose.

Salaville pursued his questioning of the evil treatment inflicted on animals, refusing to judge civilization as degraded because of it and suggesting the irrelevance of the link between compassion for animals and philanthropy. After evoking the systems of Buffon and Condillac, he seemed to favor the interpretation of Descartes, proof of the persistence in the very early nineteenth century of the representation of animals as automated machines, and as such not endowed with a sensitivity (*sensibilité*) comparable to that of men, unable to feel pain, and therefore unaware of the treatment that they receive. The animal cannot conceive of the injustice or the abuse to which he is the victim. Citing Descartes several times, Salaville argued that only human pity and sentimentality project onto animals the least intelligence, all the while acknowledging aversion to spectacles of animal mistreatment or performance.[58]

However, Salaville put forth a genuinely original idea, at once more political and more subversive than those advanced by Lamétherie in 1805. Adopting positions that he did not share, Salaville developed them logically. Thus, "if animals are what we are, there is not between them and us any more difference than one of degree, and if it is still true that this difference allows us to treat them as we do, why then would it not be the same within our species? Are there no differences of degrees among men?"[59] By posing this question, Salaville delivered an important insight already partly revealed by Lamétherie's social and political prejudices. The acceptance of a natural bond between animals and men is the first step that permits not only thinking about a slow progression from ape to man but the acceptance of the idea of a constitutive inequality among men. Ultimately, it allows one to justify the unjustifiable, which the republican Salaville summarizes as the enslavement of certain human beings, illustrated by Napoleon's restoration of slavery in 1802.[60]

Lamétherie nowhere explicitly defended slavery, but he made no secret of his aversion to Hottentots and Papuans, whom he considered to be on the limits of humanity, or of his revulsion for the "new" classes, the poorest of his world, whom he regarded as closer to animals than to civilized man.[61] Drawing on Descartes to support this awkward and weakly reasoned claim about an impermeable frontier between animals and humans, however, Salaville politically established a de facto equality between all men, saving them from any form of mutual bondage or domination under the pretext of designating some as stronger or more clever than others. Salaville's reflections challenged the legitimacy of the whole system of social domination that Napoleon constructed for France beginning in 1802 and developed further in the politics of the first empire. The greatest merit of Salaville's little book is to have invented, in a paradoxical manner, an inverted position. By far the less learned of the two authors under consideration, and the one who supported a position that the mass of his scientific contemporaries rejected—that is, the absolute difference between man and animals—Salaville came to support an otherwise audacious political position. Though questionable from a scientific perspective, Salaville's argument regarding the difference between man and animals permitted a political point of view that challenged social organization based on the quasi-natural weaknesses of the underprivileged classes or the natural enslavement of African or other peoples. While he explicitly denounced the position of Lamétherie and others at the moment of the birth of scientific racism, he did so without a full awareness of the stakes involved. At the beginning of a political regime committed to the death of liberty, Salaville defended the integrity of everyone by advocating equality for all. Yet the cost of this strategy was to leave unanswered, and for a long time, the question of animal rights.

Calling the Wild

HARRIET RITVO

Chapter SIX

When Byron wrote that "the Assyrian came down like the wolf on the fold" ("The Destruction of Sennacherib," 1815), his audience had no trouble understanding the simile or feeling its force, even though wolves had not threatened most British flocks since the Wars of the Roses. Almost two centuries later, expressions such as "the wolf is at the door" remain evocative, although the Anglophone experience of wolves has diminished still further. For most of us, they are only to be encountered (if at all) in zoos, or in establishments like Wolf Hollow, which is located in Ipswich, just north of Boston, where a pack of gray wolves lives a sheltered suburban existence behind a high chain-link fence. Their captivity has produced some modification of their nomadic habits and their fierce independent dispositions. (The pack was established twenty years ago with pups, so that only inherent inclinations needed to be modified, not confirmed behaviors.) Their relationship with their caretakers seems affectionate and playful, sometimes even engagingly doglike—so much so that visitors need to be warned that it would be very dangerous for strangers to presume on this superficial affability. The animals themselves give occasional indications that they retain the capacities of their free-roaming relatives—that though apparently reconciled to confinement, they are far from tame. When large, loud vehicles rumble past on nearby Route 133, the wolves tend to howl. And despite their secure enclosure within the built-up landscape of North American sprawl, their calls evoke the eerie menace that has echoed immemorially through the wild woods of fairy tale and fable.

The symbolic resonance of large, ferocious wild animals—the traditional representatives of what seems most threatening about the natural world—has

An earlier version of this chapter, titled "Beasts in the Jungle (or Wherever)," appeared in *Daedalus* 137 (Spring 2008): 22–30.

thus proved much more durable than their physical presence. Indeed, their absence has often had equal and opposite figurative force. The extermination of wolves in Great Britain, along with such other unruly creatures as bears and wild boars, was routinely adduced as evidence of the triumph of insular (as opposed to continental) civilization in the early modern period. As they dispersed around the globe, British settlers and colonizers set themselves parallel physical and metaphorical challenges, conflating the elimination of dangerous animals with the imposition of political and military order. In North America, hunters could claim bounties for killing wolves from the seventeenth century into the twentieth, although by the latter period wolves had abandoned most of their historic range, persisting only in remote mountains, forests, and tundras. In Africa and (especially) Asia, imperial officials celebrated the "extermination of wild beasts" as one of "the undoubted advantages . . . derived from British rule."[1]

Very occasionally, large aggressive predators could symbolize help rather than hindrance. They served as totems for people whose own inclinations were conventionally wolfish or leonine. And alongside the legendary and historical accounts of big bad wolves, a minority tradition persisted that emphasized cooperation rather than competition. From this perspective, the similarities of wolf society to that of humans implicitly opened the possibility of individual exchange and adoption. A slender line of imagined lupine nurturers ran from the foster mother of Romulus and Remus to Akela, who protects and mentors Mowgli in *The Jungle Book* (1894). But in this way as in others, Kipling's animal polity looked toward the past rather than the future. By the late nineteenth century, human opinions of wolves and their ilk had indeed become noticeably mixed. The cause of this amelioration, however, was not an altered understanding of lupine character or an increased appreciation of the possibilities of anthropolupine cooperation, but rather a revised estimation of the very qualities that had made wolves traditional objects of fear and loathing.

The shift in European aesthetic sensibility that transformed rugged mountains into objects of admiration rather than disgust is a commonplace of the history of art. For example, in the early eighteenth century, even the relatively modest heights of what was to become known as the English Lake District impressed Daniel Defoe as "eminent only for being the wildest, most barren and frightful of any that I have passed over in England, or even in Wales itself."[2] The increasingly romantic tourists who followed him gradually learned to appreciate this harsh, dramatic landscape, so that a century later the noted literary opium eater Thomas De Quincey could characterize the vistas that had horrified Defoe as a "paradise of virgin beauty."[3] Of course, this altered perception had complex roots, but it is suggestive that it coincided with improvements in transportation and other aspects of the infrastructure of tourism. As economic and technological developments made the world seem safer and more comfortable, it became possible to experience some of its extremes as thrilling rather than terrifying. Or, to put it another way, as nature began to seem a less overwhelming opponent, the valence of its traditional symbols began to change. Ultimately (much later, after their populations and geographic ranges had been radically reduced), even wild predators began to benefit from this reevaluation. The ferocity and danger associated with wolves and their figurative ilk became a source of glamour, evoking admiration and sympathy from a wide range of people who were unlikely ever to encounter them. As representatives of the unsettled landscapes in which they had managed to survive, they inspired nostalgia rather than antagonism.

Symbolic shifts were supplemented by shifts in scientific understanding, which redefined high-end predators as a necessary element of many natural ecosystems. Late nineteenth-century attempts at wild animal protection were modeled on the hunting preserves of European and Asian elites. Thus the immediate antecedents of modern wildlife sanctuaries and national parks were designed to protect individual species that were identified as both desirable (whether intrinsically or as game) and in danger of extinction, whether the bison in North America or the giraffe in Africa. They were much less concerned with preserving the surrounding web of life. In most cases, indeed, early wildlife management policies had the opposite effect. Although not all of the species targeted for protection provided conventional hunting trophies—for example, by the end of the nineteenth century, many great ape populations received some form of protection—all were herbivores. Further, none offered significant resistance to human domination of their territory. (If they did, policies could be reversed. For example, hippopotamuses, which enjoyed protection in some parts of southern Africa, were slaughtered with official encouragement in Uganda, where their belligerent attitude toward river traffic interfered with trade.)[4] Predators inclined to kill the species designated for protection received no protection themselves, either physical or legal. On the contrary, in many settings people simply replaced large predators at the top of the food chain and showed no mercy to their supplanted rivals.

Deep ancient roots can be unearthed for holistic or ecological thinking. Although most of the British pioneers of game preservation had enjoyed the classical education prescribed for privileged Victorian boys, the works of Charles Darwin may have offered more readily accessible arguments for understanding biological assemblages as interconnected wholes. Darwin provided many illustrations of the subtle and complex relationships among the organisms that shared a given territory. For example, in *On the Origin of Species* he explained the abundance of several species of wildflowers in southern England as a function of the number of domestic cats kept in nearby villages. The cats had no direct interest in the flowers, but more cats meant fewer field mice, which preyed on beehives—and fewer mice meant more bees to fertilize the flowers.[5] Nevertheless, it was not until the second half of the twentieth century that individual species were routinely considered as components of larger systems by wildlife managers, and that the standard unit of management became the ecosystem rather than the species. In consequence, large predators were redefined as essential components (even indicators) of a healthy environment rather than blots on the landscape. They often began to receive legal protection, however belated and ineffective. And there has been a movement to reintroduce them to areas that have been ostensibly preserved in their wild form or that are in process of restoration. Thus in recent decades wolves have reoccupied several of their former habitats in the western United States, both as a result of carefully coordinated reintroduction by humans, as in Yellowstone National Park, and as a result of independent (but unimpeded) migration from Canada. It is interesting that the reemergence or even the prospective reemergence of the wolf has inspired a parallel reemergence of traditional fear and hostility among neighboring human populations.

I have been using several terms as if their meanings were clear and definite, when in fact they are contested and ambiguous. As has often been noted, the cultural critic Raymond Williams characterized "nature" as "perhaps the most complex word in the English language."[6] The term "wilderness" is similarly problematic. In the context of preservation or restoration, it often collocates with words like "pristine" and "untouched," and therefore connotes a condition at once primeval and static. This connotation suggests that the first task of landscape stewards is to identify this ur-condition, but even a moderately long chronological perspective suggests that any such effort is bound to be quixotic. The environment in which modern animals have evolved has never been stable. Less than twenty thousand years ago much of North America and Eurasia was covered by glaciers. After their gradual release from the burden of ice and water, most northern lands continued to experience significant shifts in topography and climate, and therefore in flora and fauna. These natural changes have been supplemented for thousands of years by the impact of human activities. The theoretical and political problems presented by "wilderness" are knottier still. In a groundbreaking essay published more than a decade ago, William Cronon argued that wilderness and civilization (or "garden") were not mutually exclusive opposites but rather formed part of a single continuum. Far from being absolute, "the one place on earth that stands apart from humanity," wilderness was itself "a quite profoundly human creation."[7] Cronon's formulation sparked (and continues to spark) agonized resistance on the part of environmentalists who base their commitment on the notion of untouched nature.

If wildness in landscape has been effectively (if controversially) problematized, the same cannot be said for wildness in animals. The *Oxford English Dictionary* defines the adjective "wild" unambiguously, and it emphasizes its zoological application. The first sense refers to animals: "Living in a state of nature; not tame, not domesticated: opp. to TAME." In a standard lexicographical ploy, "tame" is defined with equal confidence and complete circularity as (also the first sense) "reclaimed from the wild state; brought under the control and care of man; domestic; domesticated. (Opp. to *wild*.)" But outside the dictionary these terms are harder to pin down, and their interrelationships are more complex. Like Cronon's wilderness and garden, the wild and the tamed or domesticated exist along a continuum. In a world where human environmental influence extends to the highest latitudes and the deepest seas, few animal lives remain untouched by it. At least in this sense, therefore, few can be said to be completely wild—for example, it would be difficult so to characterize the wolves that were captured, sedated, airlifted to Yellowstone, and then kept in "acclimatization pens" to help them adapt to their new companions and surroundings. And as the valence of the wild has increased, and as its definition has become more obviously a matter of assertion than of description, the boundaries of domestication have also blurred.

Not that they were ever especially clear. As twenty-first-century wolves belong to a long line of animals whose wildness has been compromised, tameness has conversely also existed on a sliding scale. According to the *OED*, both "wild" and "tame" have persisted for a millennium, remaining constant in form as well as core meaning, while the language around them has mutated beyond easy comprehension, if not beyond recognition. But this robustness on the level of abstraction has cloaked imprecision and ambiguity on the level of application or reference. Although medieval farmers and hunters may have had no trouble distinguishing livestock animals from game or vermin, it would have been difficult to extract any general definition from their practices. The impact of domestication varied from kind to kind, as well as from creature to creature. The innate aggression of the falcons and ferrets who assisted human hunters was merely channeled, not transformed; when they were not working, they were confined like wild animals in menageries. Then as now, people exerted much greater sway over their dogs than over their cats, who were mostly allowed to follow their own instincts with regard to rodents and reproduction. Medieval cattle, the providers of labor as well as of meat, milk, and hides, led more constrained lives than did contemporary sheep, and pigs were often left to forage in the woods like the wild boars that they closely resembled.

22 *Head of the Chillingham Bull Shot by H.R.H. the Prince of Wales,* 1879. Engraving. From John Storer, *The Wild White Cattle of Great Britain* (London: Cassell, Petter and Galpin, 1879), 169.

With hindsight, even these relatively tame cattle could appear undomesticated, especially as wildness gained in glamour. Thus changes in the animals' physical circumstances were complicated by changes in the way they were perceived. In the late eighteenth century, for example, a few small herds of unruly white cattle, who roamed like deer through the parks of their wealthy owners, were celebrated as aboriginal and wild. As the Earl of Tankerville, whose Chillingham herd was the most famous, put it, his "wild cattle" were "the ancient breed of the island, inclosed long since within the boundary of the park."[8] The "ancient breed" was sometimes alleged to be the mighty aurochs (the extinct wild ancestor of all domestic cattle, which had been eliminated in Britain by Bronze Age hunters; the last one died in Poland in the seventeenth century), which gave these herds an ancestry distinct from that of ordinary domestic cattle. To increase or underscore their distinctiveness, the white cattle were never milked, and if their meat was required for such ceremonial occasions as the coming of age of a human heir, they were hunted and shot, not ignominiously slaughtered (fig. 22). Throughout the nineteenth century, their autochthonic nobility continued to inspire the effusions of such distinguished poets and painters as Sir Walter Scott and Sir Edwin Landseer, as well as the expenditures of newly wealthy landowners eager to bask by association in the prestige of wild nobility and ancient descent.

But even at the height of their renown, it was clear that these claims to wildness included a large measure of wishful thinking. Skeptics wondered persuasively whether, even assuming that the nineteenth-century emparked herds lived in a state of nature, that state represented a historical constant or a relatively modern restoration.[9] Many who investigated the background of the herds concluded that they were feral at best (at wildest, in other words)—that they were the descendants of domesticated animals, whether originally owned by Roman settlers or by later farmers. Modern anatomical and genetic research has confirmed these doubts, firmly connecting the emparked herds with the ordinary domestic cattle of the medieval period.[10] But so great is the continuing appeal of wildness, and so limited the persuasive force of scientific evidence, that a recent president of the Chillingham Wild Cattle Association has nevertheless asserted that "although there is still much that is not known about the origins of the Chillingham Wild Cattle, one fact that is certain is that they were never domesticated."[11]

Only a few people possessed the resources necessary to express their admiration for the wild, and their somewhat paradoxical desire to encompass it within the domestic sphere, on such a grand scale. But numerous alternative options emerged for those with more restricted acres and purses. An increasing variety of exotic animals stocked private menageries. The largest of these were on a sufficiently grand scale to have also included a cattle herd, if their owners had been so inclined—for example, those of George III or the thirteenth Earl of Derby, which accommodated large animals like kangaroos, cheetahs, zebras, and antelopes. Smaller animals required more modest quarters, and parrots, monkeys, canaries, and even the celebrated but ill-fated wombats owned by the poet Dante Gabriel Rossetti could be treated as pets. Breeders attempted to enhance or invigorate their livestock with infusions of exotic blood. If they were disinclined or unable to maintain their own wild sire, they could, in the 1820s and 1830s, pay a stud fee to the newly established Zoological Society of London for the services of a zebu or a zebra. In Australia, Russia, Algeria, and the United States, as well as in Britain and France, the acclimatization societies of the late nineteenth century targeted an impressive range of species for transportation and domestication, from the predictable (exotic deer and wild sheep) to the more imaginative (yaks, camels, and tapirs).[12] So difficult (or undesirable) had it become to distinguish between wild animals and tame ones that exotic breeds of domestic dogs were exhibited in Victorian zoos, and small wild felines were exhibited in some early cat shows.

Even human beings were the persistent subjects of similar confusion. In both his correspondence and his lectures at Uppsala University, the preeminent eighteenth-century systematizer Linnaeus repeatedly suggested that he found it difficult to distinguish clearly between humans and apes.[13] Not only did he devise the primate order to accommodate humans, apes, monkeys, lemurs, and bats, but he included one incontestably wild species within the human genus: the sole occupant of *Homo troglodytes* was the orangutan.[14] The evidence offered by this placement is, however, ambiguous. The orangutan was also known to Linnaeus's contemporaries as *Homo sylvestris,* or the wild man of the woods (a translation from Malay, though not of the Malay word for the orangutan), and, at a time when the unity of the human species was the subject of vigorous debate, there was widespread uncertainty among naturalists about whether or not orangutans were human.

Among the general public, the possibility that apes might actually be people lingered in various ways. The illustrations in books of popular natural history often portrayed apes as particularly human in both appearance and behavior, showing them assuming erect posture, using human tools (frequently a walking stick), and approximating human proportions in the torso and limbs (fig. 23).[15] This visual convention was not confined to the page or the canvas. It was also constantly reenacted in the displays of the chimpanzees and orangutans that constituted popular components of nineteenth-century zoos and menageries, featuring apes that ate with table utensils, sipped tea from cups, and slept under blankets. Consul, a young chimpanzee who lived in Manchester's Belle Vue Zoological Gardens at the end of the nineteenth century, greeted the public dressed in a jacket and straw hat, smoked cigarettes, and drank his liquor from a glass.[16]

In addition, rumor persistently whispered that these visual analogies might represent more substantial and productive connections. Thus one seventeenth-century report featured a "poor miserable fellow" who had copulated with a monkey "not out of any evil intention . . . , but only to procreat a Monster, with which he might win his bread."[17] At the end of the eighteenth century the surgeon and naturalist Charles White reported that orangutans had "been known to carry off negro-boys, girls and even women . . . as objects of brutal passion"; more than sixty years later the Anthropological Society republished Johann Friedrich Blumenbach's summary of travelers' accounts that "lascivious male apes attack women."[18] Still more suggestively, White recorded rumors "that women have had offspring from such connection" and proposed that, "supposing it to be true, it would be an object of inquiry, whether such offspring would propagate, or prove to be mules."[19] Blumenbach, more cautious, asserted, "that

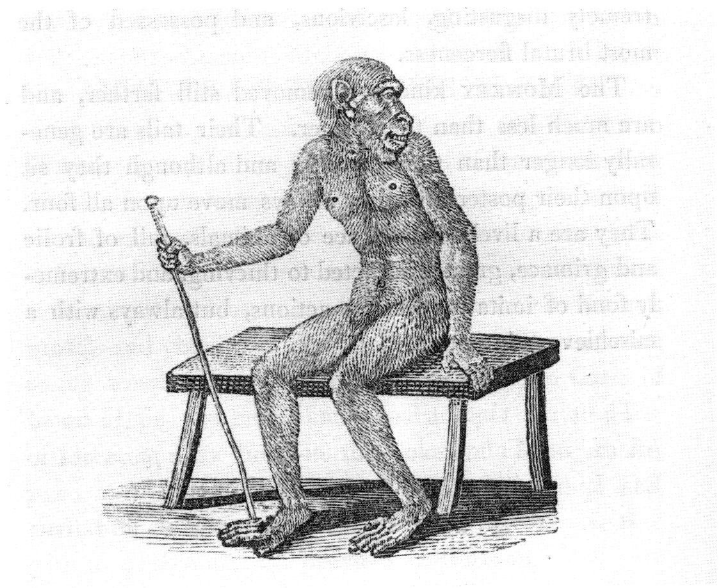

23 Thomas Bewick, *The Oran-outang, or Wild Man of the Woods,* [1790]. Engraving. From Bewick, *A General History of Quadrupeds* (1790) (Newcastle: T. Bewick, 1824), 452.

such a monstrous connection has any where ever been fruitful there is no well-established instance to prove."[20]

Outside the community of experts, claims could be less restrained, or more enthusiastic. For example, a Victorian impresario advertised the merely hairy Julia Pastrana as "a hybrid, wherein the nature of woman predominates over the ourang-outangs."[21] Because the evolutionary theories that gained currency in the late Victorian period assumed the existence of extinct forms intermediate between humans and apes, at least in the sense of having given rise to both modern groups, the rhetoric of evolution could also be deployed to suggest that human-ape mixtures existed in the present, as well as in the ancestral past. For example, a Laotian girl named Krao was exhibited in 1883 as "Darwin's missing link," not only because she was unusually hairy but because she allegedly possessed prehensile feet and could pout like a chimpanzee.[22]

The popular appeal of wild animals, however defined, has continued to increase as such creatures have become more accessible, either in the flesh or in the media. So entangled have wildness and domesticity become that it is now necessary to warn visitors to North American parks that roadside bears may bite the hands that feed them, and it is now possible for domesticated animals to represent nature. This extended symbolic reach was demonstrated in 2001, when foot-and-mouth disease struck British livestock. Because the disease spreads rapidly and easily, the government prescribed a cull not only of all infected herds and flocks but of all apparently healthy livestock living in their vicinity. Although outbreaks were widespread, the greatest number of cases occurred in the Lake District, the starkly dramatic landscape that had been disparaged by Daniel Defoe and praised by Thomas De Quincey; it is now the site of England's largest national park. Video and print coverage of the cull, which took the spectacular form of soldiers shooting flocks of sheep and then immolating them in enormous pyres, thus featured some of the nation's most cherished countryside as background.

The ovine victims also had iconic status. Most of them belonged to the local Herdwick breed, and at first the intensive cull seemed to threaten its very survival. What was at stake was not merely adaptation to a demanding environment, since several other British hill breeds look very like the Herdwicks and share their physical and emotional toughness. The Herdwicks' special claim to consideration was their connection to their native ground, itself a kind of national sacred space. Not only were the sheep acknowledged to possess detailed topographical information about the hills they inhabited, but their owners claimed that they transmitted it mystically down the generations, from ewe to lamb. So well recognized was their attachment to their home territories that when a farm was sold, the resident Herdwicks were conventionally included in the bargain, on the theory that if they were taken away, they would soon manage to return. And despite strong historical indications that the ancestral Herdwicks had arrived in the vicinity of the Lake District by boat, and the further fact that all British sheep descend from wild mouflons originally domesticated in the eastern Mediterranean region, they were traditionally celebrated as indigenous, "peculiar to that high, exposed, rocky, mountainous district."[23] An article in the preeminent Victorian agricultural journal asserted that the Herdwicks possessed "more of the characters of an original race than any other in the county" and that they showed "no marks of kindred with any other race."[24] Twenty-first-century journalists reporting on the threatened toll of foot-and-mouth disease adopted similar rhetoric.

As the sheep were nativized, they were also naturalized. A reporter for the *Independent* feared that if the Herdwicks disappeared, the whole ecology of the region might be changed "beyond recognition."[25] And, since the dramatic bare uplands of the Lake District have been maintained by nibbling flocks for at least a millennium, his concern was not completely unreasonable. Thus, whether technically indigenous or not, and although they are incontestably domesticated, the Herdwicks have become compelling symbols of the apparently untamed landscape they inhabit—more compelling than the numerous wild birds and

24 *The Champion Herdwick Ram, "Lord Paramount," 1882.* From Frank W. Garnett, *Westmorland Agriculture, 1800–1900* (Kendal, England: Titus Wilson, 1912), facing p. 154.

small mammals with whom they share it. Like the landscape itself, they seem wilder than they are; that is, they appear to be independent and free ranging, but their lives (and indeed their very existence) are ultimately determined by human economic exigencies. They are both accessible (that is, there are a lot of them and they are everywhere, not only in the fields but grazing and napping beside the roads and even on top of them) and also inaccessible (that is, they are skittish and tend to retreat when approached). The armed assault on the Herdwick sheep was therefore perceived as an attack on both the domesticated countryside and the unspoiled natural landscape. In both the sheep and their environment, the wild and the tame had inextricably merged.

If vernacular usage illustrates the increasing slippage between wildness and tameness in animals, scientific classification has made a similar point from the opposite direction. The species concept has a long and vexed history. The study of natural history (or botany and zoology) requires that individual kinds be labeled, but for many plants and animals (those that, unlike giraffes, for example, have very similar relatives) it has been difficult for naturalists to tell where one kind ends and the next begins. Darwin's theory of evolution by natural selection provided a theoretical reason for this difficulty, and his shrewd observations that "it is in the best-known countries that we find the greatest number of forms of doubtful value" and that "if any animal or plant . . . be highly useful to man . . . varieties of it will almost universally be found recorded" offered a more pragmatic explanation.[26] The classification of domesticated animals has epitomized this problem. That is, none of them has become sufficiently different from its wild ancestor to preclude the production of fertile offspring (the conventional if perennially problematic definition of the line between species), and some mate happily with more distant relatives. Nineteenth-century zookeepers enjoyed experimenting along these lines, and zoo goers admired the resulting hybrids between horses and zebras, domestic cattle and bison, and dogs and wolves.[27]

Despite these persuasive demonstrations of kinship, however, from the eighteenth-century emergence of modern taxonomy, classifiers have ordinarily allotted each type of domestic animal its own species name. While recognizing the theoretical difficulties thus produced, most modern taxonomists have continued to follow conventional practice. Domestic sheep are still classified as *Ovis aries,* while the mouflon is *Ovis orientalis,* and dogs as *Canis familiaris,* while the wolf is *Canis lupus.* The archaeozoologist Juliet Clutton-Brock explains this practice as efficient (it would be unnecessarily confusing to alter widely accepted nomenclature) as well as scientifically grounded, at least to some extent (most domestic animal populations are reproductively isolated from wild ones by human strictures, if not by biological ones).[28] But it also constitutes a simultaneous acknowledgment of the artificiality of the distinction between wild animals and domesticated ones, and of its importance and power. Vernacular understandings can trump those based on anatomy and physiology.

The implications of making or not making such distinctions extend beyond the intellectual realm. They construct the physical world as well as describe it. Although the howls of the wolf may retain their primordial menace, the wolves who make them have long since vanished from most of their vast original range, and they are threatened in much of their remaining territory. To persist or to return, they need human protection, not only physical but legal and taxonomic. With the advent of DNA analysis in recent decades, the taxonomic stakes have risen, so that even animals that look and act wild may be found genetically unworthy. Thus efforts to preserve the red wolf, which originally ranged across the southeastern United States, have been complicated by suggestions that it is not a separate species but a hybrid of the gray wolf and the coyote. No such aspersions have been cast upon the pedigree of the gray wolf, but nevertheless every attempted gray wolf restoration has triggered human resistance, and local challenges to their endangered status inevitably follow even moderate success. If domestic dogs were returned to their ancestral taxon, wolves would become one of the commonest animals in the lower forty-eight states, rather than one of the rarest. Their survival as wild animals depends on the dog's continuing definition as domesticated.

Trophies and Taxidermy

NIGEL ROTHFELS

Chapter SEVEN

There are elephants on Kenia that have never lain down for a hundred years.

—CARL AKELEY

Whatever else one might take from it, those who have seen Douglas Gordon's installation *Play Dead; Real Time,* originally shown in the Gagosian Gallery in New York in 2003 (fig. 25), have probably considered the work to be *also* about elephants. The piece consists of video of a trained elephant, Minnie, shown from various perspectives and displayed on two large screens and a monitor. While the smaller monitor focuses on Minnie's eyes and other shots of her body, the large screens show her standing up or lying down on her side, remaining motionless as if dead in the white space of the Chelsea gallery. After lying on her side for some time, Minnie suddenly lifts her head, balancing the effort by simultaneously lifting her rear slightly off the floor. With this, Minnie begins a complex series of three heaving movements to get up. After this first move, Minnie's head then goes back down to the floor while she uses her shoulder as a fulcrum to lift the lower part of her body higher up. The rocking motion then quickly shifts in the other direction as the rear is brought down quickly to the floor, allowing Minnie to bring up her head and shoulders and put her front legs underneath her. From here, she is able to stand up on her front legs and rotate her back legs underneath her, kneel on one rear leg while standing up on the other, and then finally come to a complete standing position.

25 Douglas Gordon, *Play Dead; Real Time,* 22 February–29 March 2003. Installation view, Gagosian Gallery, New York. Artwork © Douglas Gordon. Courtesy of Gagosian Gallery and Studio lost but found.

Play Dead; Real Time can be interpreted in many ways, but it is at least partly a stunning exploration of scale and proportion, of the sheer size and physicality of an elephant. More than this, though, it is about the arresting image of an elephant appearing dead, holding the trick of "play dead" for so long that the viewer is forced to move beyond ideas of "play." Tapping into a history of grotesque images of elephants, killed for their tusks or dead from starvation and from management culls, the work reminds viewers of the language and realities of death and extinction that have haunted thoughts about elephants for more than a century. As much as this work is about death, however, it is not about the rotting, hacked-up remains of elephants in the African bush—the dominant images of elephant death in the 1970s and '80s. The experience of this work is more like quiet loss and mourning—feelings that have become increasingly associated with elephants in both popular and specialist contexts in recent decades. Whether one is studying the results of field research by people like Iain Douglas-Hamilton or Cynthia Moss, reading an imaginative work like Barbara Gowdy's *The White Bone,* or just talking to those elephant keepers at zoos who have become convinced that it is critical, when an elephant dies, to let the other elephants in the group spend time with the body of the deceased, it is clear that when Minnie plays dead in the quiet gallery space, the death she is describing is not a death by a trophy hunter, a death by ivory hunters carrying machine guns, or even a death by age or disease. This death is somehow both less specific and more intellectual or, perhaps, spiritual; it seems to be about the passing away of the very possibility of something like an elephant.

The press release for the show notes that "the elephant has classically symbolized memory, and here functions as a trope for our own remembrances of circuses, zoos, and nature documentaries, various situations where the chaotic power of the wild is held safely at a distance and is controlled." Highlighting the human contexts in which elephants live, *Play Dead; Real Time* uses a classic animal trick to examine critical ideas of art, performance, memory, representation, and elephants. As much as Gordon's piece echoes longer memories of elephant lives and deaths, though, it also recasts those scenes in a twenty-first-century tone. This chapter traces those continuities and contrasts through an examination of trophy shots in the late nineteenth and early twentieth centuries. By showing how these shots are both part of longer-lasting ideas about hunting but also historically distinctive, this chapter places the work of Douglas Gordon and other contemporary views of animals—elephants in particular—in a necessary historical perspective.

Elephants are among the animals that are able to sleep while standing up. Indeed, in the case of those who have lived for many years in zoological gardens, it is not too uncommon to find one that has not lain down for years. This tidbit of natural history about elephant behavior reaches back millennia. Some accounts explained the phenomenon by speculating that elephants lacked leg joints—an idea that could still be found well into the eighteenth century. The late antique *Physiologus,* for example, noted that elephant hunters would cut almost completely through particular forest trees that the animals used as supports when sleeping. When an elephant returned to and leaned up against its sleeping tree, both the tree and the elephant would topple over, and because the elephant couldn't get back up, the hunters would simply return at their leisure to kill the beast.

It is not too difficult to imagine the origin of an idea like elephants with jointless legs—it is more difficult to comprehend the physics of how an animal as big as an elephant could ever get up from the ground, or even lie down in the first place. There is also something quite unusual about the shape of an elephant when it actually does lie down. Most of the weight of an eight- to twelve-thousand-pound elephant is in the gut and huge digestive system, and when an elephant lies on her side, the rib cage shifts up to accommodate her bulk. From the end of her tail, then, the line of her body rises smoothly and sharply to a high point just rear of midway down her body; the line then descends again, more gradually this time, past the shoulders to the head and the extended trunk. No other animal lying on its side looks quite like this "mountain of flesh"—as hunters often described it in the nineteenth century.

Once we understand that adult elephants are rarely seen lying down, and that when they do lie down their shape changes radically, we can begin to make more sense of the innumerable images of hunters standing on, sitting on, or just leaning up against dead elephants. Consider, for example, the illustration by Sydney P. Hall of the Prince of Wales standing on the elephant he shot in Ceylon in 1875 (fig. 26). In the illustration *Ceylon—The Dead Elephant,* the animal is portrayed lying on her right side. As one of the professional elephant hunters engaged by the prince examines her eye (slapping the eye was a common test to see if the animal was actually dead), the prince stands with his left foot anchored on the solid left thigh of the animal, while his bent right leg rests further up the elephant's back. The prince's weight is on his lower leg, and his stance suggests someone looking out from a summit over a valley. The prince has his rifle propped nonchalantly over his right shoulder, while his left hand is in the waist pocket of his hunting jacket. Hall portrays the prince as a calm center surrounded by the jubilation of his hunting companions and scattered native beaters. While the other hunters whoop, cheer, and salute by taking off their hats, the prince, the only European to have lost his hat in the midst of the hunt (it can be seen upturned in the foreground), echoes the apparent lack of enthusiasm, in deference to his solemn dignity, that we see in classic equestrian statues depicting imperturbable generals perfectly balanced atop powerful and violent horses.

Hall was a member of the prince's hunting entourage, and, it is claimed, his original illustration was made on the spot. Also in the group was the prince's diarist for his trip to the East, William Howard Russell, who describes the events leading up to the prince's standing atop his elephant. Preparations for the event had been going on for several weeks prior to the prince's arrival, as structures were built in the forest through which drivers would be able to move a herd of elephants within shooting range of the prince. On the morning of the hunt, more than a thousand beaters and several experienced European elephant hunters began to move several groups of elephants toward the prince's position. After hours of waiting and listening to the sounds of the beaters and gunshots in the distance, suddenly "branches crashed, and trees shook violently, a couple of

26 Sydney P. Hall, *Ceylon—The Dead Elephant,* 1875. Engraving. From William Howard Russell, *The Prince of Wales' Tour: A Diary of India* (New York: Lovell, Adam, Wesson, 1877), facing p. 254.

shots were heard—an elephant rushed, like some great rock, down the hillside within twenty yards of the Prince, who fired." According to Russell, the prince hit the large bull but the animal disappeared into the forest. After a few more minutes, one of the hunters ran up to the prince and said, "Sir! If you will come with me I think I can get you a shot. I have wounded an elephant; I know where he is, and you can kill him." The prince immediately came down from his post with several of his companions and the hunt proceeded on foot. Russell writes that the "heat was great; it was impossible to see two yards ahead. Shooting hats were lost, clothes torn. Suddenly the elephant which had been wounded was discovered through the jungle." The prince fired at this second elephant and it "dropped at once, and lay as if dead." According to Russell, Hall began to sketch this animal, while the prince, the other hunters, and a group of beaters pursued other elephants. Apparently, though, the elephant Hall was sketching soon got to his feet, the surprised Hall retreated, and this second wounded elephant also disappeared into the jungle.

In the meantime, the hunting party had come to a halt and become uneasy upon realizing that they were very close to elephants they couldn't actually see.

"At any moment," writes Russell, "an elephant might rush out; evasion and escape were hopeless, for in such a jungle no man could do more than very slowly creep, whilst the elephant could go through the brush as a ship cleaves the water." Then, suddenly, an elephant charged the group; the prince saw it and fired at ten yards. This third elephant, though, also disappeared into the jungle. According to Russell, "the huntsmen continued in pursuit cautiously, but the creepers and the thick under growth made stout resistance, so that their progress was slow, and not unexhausting. In a few minutes more another elephant was seen, where the bush was not so dense, by the side of the rivulet. The Prince took deliberate aim and fired. The great beast toppled, and fell over on its side in the stream, where it dammed up the waters!" The prince climbed down the bank and, "assisted by the hunters, got into the water and climbed upon the inert mountain of flesh." All of the European hunters and a few beaters assembled around the prince. They "cheered again and again, and the whole party whooped and woke up the glade with their cries, as the Prince was seen standing on the prostrate body—which was not that of the redoubtable tusker." This fourth elephant, the only one the prince killed immediately, turned out, as Russell notes obliquely, not to be a large male "tusker," and thus more worthy game, but a female, and the "Prince, according to custom, cut off the tail." The Singhalese beaters tore pieces from the ears as "trophies," and the prince, in Russell's depiction if not Hall's, "was streaming with perspiration, his clothes were wet, and torn to shreds."[1]

Hall's and Russell's accounts were intended to create an image of the Prince of Wales as one of the most accomplished, powerful, brave, and wise men of the world. Brought along to make an official record of the expedition, these men were essentially hired to testify to the prince's awe-inspiring qualities, but the audience for these descriptions did not simply accept the accounts as unvarnished. That the narratives of the "great hunters" of the nineteenth century proclaimed the bravery of the men and the danger of the game does not mean that the hunters were universally understood as heroic. In fact, big-game hunting was controversial in the nineteenth century, and while it is one thing to describe the Prince of Wales as a great lord of the hunt, at least some observers in Ceylon, and some readers back home, surely found the whole business somehow ridiculous.[2] A report about the prince's trip in the July 1877 edition of *Appleton's Journal: A Monthly Miscellany of Popular Literature* makes this point clear, if subtly. The report is based on Russell's narrative and presents the key events in the official account with a sort of winking knowingness. The author is addressing an audience familiar with the celebrity of the Prince of Wales, an audience that could see an account of the perilous adventures of European royalty visiting "their lands" in the East as puffery. The tone of the article is clear in the opening sentence: "It was every way fitting that 'The Most High, Most Puissant, and Most Illustrious Prince, Albert Edward, Prince of Wales,' should visit India and see something of the two hundred and odd millions of people who have since 1858 come to look upon him as their future 'Shahzadah,' or whatever other Oriental designation should be held to be the equivalent of 'Emperor of India.'" That the author misuses the term *shahzadah*—the prince was *already* the *shahzadah,* the heir apparent—only contributes to the overall half-snickering feel of the article. The description of the prince makes this even more apparent. "Despite his long string of titles, beginning with 'Duke of Saxony, Duke of Cornwall and Rothsay, Earl of Chester, Carrick, and Dublin, Baron of Renfrew, Lord of the Isles,' knight of all sorts of orders, field-marshall in the army, and colonel of three regiments, and running down to 'Barrister-at-Law, D.C.L., LL.D., etc.,' the Prince can hardly be looked upon as an heroic personage. He is a middle-sized, rather pudgy gentleman of six-and-thirty, quite nice-looking, with a noticeable thinness of hair at the top of his head."[3]

When the author turns to the events of the elephant hunt, his emphasis is, tellingly, just slightly different from that of Russell. He briefly describes the prince's last two shots:

> On they crept through the jungle, the noise of invisible elephants being heard close at hand. Suddenly one came charging straight for them. The prince fired at ten yards' range, but apparently missed, for the elephant disappeared in the jungle. In a few moments another was perceived in a spot where the less dense thicket gave a chance for deliberate aim. The prince fired, and the huge beast fell upon its side and toppled into the little stream. They crept up to it and found it dead. The victorious marksman waded into the shallow water, and was boosted up upon the carcass, a most perspiring, ragged, hatless, yet triumphant royal personage. In Ceylon only the male elephant has tusks, and this was not a "tusker"; so that the only trophy was the tail, which the prince cut off, carried away, and, for aught we are told, took home with him to England.[4]

Spilling no ink describing the jungle as deadly and otherworldly, the hunt as full of thrill and suspense, or the terrifying moment when the hunters realize that they have become the hunted, the author of the account for *Appleton's* instead points to the banality of the hunt. In this version, the prince misses an elephant at ten yards, and when he finally does kill one, the author avoids descriptions of the cheering throng and concludes simply, "They crept up to it and found it dead. The victorious marksman waded into the shallow water, and was boosted up upon the carcass, a most perspiring, ragged, hatless, yet triumphant royal personage."

In a way, both the official account of the prince's trip and the *Appleton's* version work with stock characters. In Russell's account, the prince plays the role of the restrained and awe-inspiring leader; in the *Appleton's* piece, the prince is essentially a buffoon. These characters existed both well before and after the actual trip in 1875–76. The "prince as buffoon hunting an elephant," for example, is immediately recognizable in a six-frame cartoon published a decade earlier, in the April 1868 issue of *Ballou's Monthly Magazine.* Titled "Prince Alfred and His Elephant Hunt," the six comical illustrations mark the key events of a royal elephant-hunting trip in southern Africa. In the first illustration, two Europeans, shown facing each other, hoist bottles with their right hands while holding the muzzles of guns at their sides. Between them, facing the reader, an African woman, wearing only a small cloth around her waist, holds the muzzle of a third gun before her in both hands as the stock rests on the ground. The caption reads, "His royal highness Prince Alfred, after landing at the Cape of Good Hope, drinks success to his elephant hunt." The next frame depicts six young elephants playing something like a game of leapfrog while two adult elephants look on. According to the text, "Alfred is shown a number of baby elephants at play." In the third illustration, the two Europeans, now flanked by two Africans—a tracker and a gun bearer—observe an elephant from behind, and the caption reads: "The prince obtains a rear view of a full grown elephant." This is followed by a fourth frame showing an unarmed European from behind standing spread-eagled before three charging elephants. The caption reads, "The prince obtains a front view of the game." The punch line appears beneath the fifth illustration, which shows an elephant chasing the Europeans and Africans up the trees: "The elephant obtains a rear view of the royal prince." The final frame shows an elephant disappearing into the woods while a European looks on approvingly at another European dancing a jig, hat doffed and in hand, on the belly of a dead elephant, which is shown lying on its back, all four feet in the air.[5]

The illustration of an elephant on its back is as unrealistic as the depiction of Prince Alfred dancing on the animal. While an elephant may look like a "mountain of flesh" when lying on its side, one would not be able to dance on the soft belly of the creature. Indeed, a dead elephant is not the sort of thing one can do anything with terribly easily. This is the core preposterousness of Hall's image of the Prince of Wales standing on the elephant, an absurdity hinted at in the *Appleton's* article when the author notes that the prince "was boosted up upon the carcass." After drawing attention to the prince's rather civilized ("pudgy") physique, the author invites us to consider the decidedly indecorous moment when the prince, probably on his hands and knees and pushed from behind, had to get himself onto the elephant for his trophy shot. When the *Appleton's* author combines the word "boosted" with "perspiring, ragged, hatless, yet triumphant," he begs us to imagine what happened before the heroic illustration, and the contrast is made even more pronounced by his emphasizing that the quarry was not a terrifying, dreadful "tusker" but a female, for which the only physical trophy would be her tail.

The illustration of the Prince of Wales was, of course, not the first time anyone had been pictured standing on top of an elephant, and it wouldn't be the last. Consider the photograph by Carl Akeley of Delia Akeley standing on the first elephant she shot, a large bull, a photograph that in many ways echoes the drawing of the prince (fig. 27). The photograph has been composed head-on to the elephant. The animal is lying on his right side. We see the arch of his left tusk, while the tip of his right tusk is just visible underneath the trunk. The shot is a dorsal perspective of the elephant, as we look down the line from his trunk to the top of his head and shoulders; the view ends with the bulge of the elephant's gut just before his hind legs. Delia is standing on his rib cage, just to the rear of the upper left leg. Her triangular stance echoes that of the prince; left leg back, right leg forward, feet at ninety degrees to each other. She is holding the muzzle of her gun in her left hand, while the stock rests near her left foot. Her right hand is, again like the prince, in the pocket of her hunting jacket. Even the clothes of the prince and Delia are essentially the same—indeed, they may have been supplied by the same London outfitters. In both images, the hunter is seen looking off into the distance.

It is important to note, though, that however essentially similar such pictures are—at base, they all depict a hunter with a dead animal—there are critical differences among them. Indeed, even though these two images are strikingly similar, they remain fundamentally different in ways well beyond the fact that one image is a drawing and the other, a photograph. The differences are suggested in the "cast of characters" of the two works. Whereas Hall wanted to show the adulation of the Europeans and Singhalese surrounding the prince, Carl has removed a professional hunter and various Africans from the scene, and has isolated Delia and the elephant, creating a double portrait. The elephant here is not the giant piece of furniture one finds in many such shots, nor is it an object of amusement or curiosity, nor, finally, is it posed as if it had *already* become a taxidermic trophy—a rug or wall mount—as can often be seen with trophy shots of deer, bears, and lions, but only occasionally with elephants when they collapse upright, as sometimes happens. Foreshortened and abstracted by the angle of the camera shot, the elephant here appears more like a giant boulder

27 Carl Akeley, *Delia Akeley,* 1905.

or tree. This elephant tells a story fundamentally about scale, and about how big things like elephants and redwoods fall to the ground at the hands of humans. Delia, by contrast, is portrayed as a hunter, and more than that, as a female hunter who has brought down the largest game. Hair in a bun tucked under her hat, wearing a costume intended for men (clothes chosen for practical reasons but that also paradoxically emphasized her gender), and holding a gun reaching almost as high as her shoulder with a comfort and surety that seems to proclaim, "I, Delia Akeley, brought down this monster," Delia's manner here is the opposite of the spectacular (or silly) bravado of the Prince of Wales. His swagger has been replaced in her with a calm resolve repeatedly admired by Carl in his books and essays—a resolve that, as she herself makes clear in her memoir, *Jungle Portraits,* she didn't always have. Her expression is not, in my view, the "semi-shock" seen by Penelope Bodry-Sanders in her insightful biography of Carl Akeley, but more a portrayal of calm concentration. To be clear, I'm not claiming anything about Delia's *actual* emotional state at that moment; I'm pointing out only that the tone of this image is absolutely consistent with the persona of Delia Akeley that Carl promulgated in his writings. The same tone is also evident in the picture of her, taken later that day, being "blooded" by Richard John Cunninghame, the professional hunter Carl had engaged as a guide.[6] While it is *possible* that this ritual was understood as particularly significant by Cunninghame and the Akeleys, we must also recognize that this photograph, like the one of Delia standing on the elephant, was not some sort of quick snapshot in the field, in which those portrayed were carrying on somehow without regard for the presence of the camera. Given the limitations of field photography at the time and Carl Akeley's repeated protests that he was not a big-game hunter but a scientific collector, we should understand these photographs—like essentially all trophy photographs, I would argue—as carefully staged to tell a story to audiences *not in the field.* These photographs are less records of what happened than claims about what happened. The critical question then becomes, "Just what is being claimed?"

In his important essay "The Cinema as Taxidermy: Carl Akeley and the Preservative Obsession," Mark Alvey argues that Akeley's lifelong quest in taxidermy, photography, filmmaking, and sculpture was to capture not simply the physical form of an animal but a deeper truth about the species and life in general. In this, Alvey agrees with Donna Haraway's point that Akeley's taxidermy was intended to reveal more about the essence of the animal than any living representative of the species could do. Haraway writes that the animals in the Akeley dioramas of the American Museum of Natural History (AMNH) "transcended mortal life" to become "a spiritual vision made possible only by their death and the literal re-presentation." Haraway describes taxidermy as "a politics of reproduction" that fulfills "the fatal desire to represent, to be whole."[7] For Alvey, though, this "hopelessly obsessive" quest must always necessarily fall short. He writes that none of Akeley's efforts "could ever actually capture the essence, the truth, the beauty, the ethos."[8] In the end, Alvey asserts,

> a mounted kudu at one level impresses, and at another disappoints—it is the animal, and yet it is not. It is both real (skin, hair, horns, DNA), and fake (*papier-mâché* internals, metal leg supports, and glass eyes). A "stuffed" panda in a diorama is more real than a photograph, and yet somehow less real than an example of the same species at the zoo. Through skin, film, clay, and bronze Akeley illuminated different aspects of these creatures, some purely physical, some less tangible. Each feature captured something of the animal's true nature; each medium and each attempt flowed back to Akeley's core fascination with nature and animals, and that founding obsession to communicate some of the truth that he knew and the passion he felt about them.[9]

For Akeley, communicating a truth meant reaching beyond the purely physical. It is clear, for example, in both his groundbreaking muskrat group at the Public Museum in Milwaukee and his *Four Seasons of the Virginia Deer* at the Field Museum in Chicago, that Akeley was trying to do much more than show animals in the most scientifically accessible way. At the same time, it is also apparent, most obviously in his bronze sculptures but also in his taxidermy, that Akeley was hoping to convey a good deal more than just the life history of the animals—the ostensible focus of these exhibits. In fact, despite what has repeatedly been seen as his compulsive desire to re-create the natural scenes exactly as they were in situ (the famous case here being that of the mountain gorilla group at the AMNH, with its intense attention to floral accuracy), when it came to telling a story (telling the truth) through his work with taxidermic animals, Akeley was more interested in portraying the animals *as he wanted them to be seen* than as he had actually seen them. Thus the gorilla and elephant groups at the AMNH had never been observed as they were portrayed, and the stunning *Fighting Bulls* at the Field Museum depicted an event that Akeley himself doubted ever really happened; he insisted, in fact, that he had "never heard or seen African elephants fighting each other. They have no enemy but man and are at peace amongst themselves."[10] So why portray the elephants fighting?

When the Akeleys returned from their collecting expedition to British East Africa in 1907, they brought back seventeen tons of skins and specimens, but just as they hadn't photographed every kill in the field, they didn't mount every animal for exhibition.[11] Select animals were chosen for mounting, and the most important of these were clearly the two elephants, *The Fighting Bulls,* completed in 1909 (fig. 28). The work is not one of Akeley's typical habitat dioramas; the elephants are not shown from a one-point perspective in a complex environment with a carefully crafted background painting seamlessly blending into a foreground reconstructed to look like the African plains. The work, in fact, bears a closer affinity with contemporary bronze sculptures of animals in dramatic poses. Here, the giant elephants are placed on a sculptural pedestal, around which the original audience in the Rotunda of the old Field Columbian Museum (originally the Fine Arts Building of the World's Columbian Exposition and now the Museum of Science and Industry) could walk as they studied the work from all sides. The two large African bull elephants are engaged in battle. The smaller of the two appears to be charging the larger, bringing his single tusk into the chest of his opponent. The stance of the smaller elephant is unnaturally stretched out, with all four feet solidly on the ground. An actual living elephant would not normally assume this exaggerated position, but Akeley seems to have chosen it to emphasize a powerful move forward, or perhaps a sort of boxing

28 Carl Akeley, *The Fighting Bulls,* 1909. Taxidermic elephants. The Field Museum, Chicago.

swing from left to right.[12] Indeed, even the elephant's skin, especially behind the left front and left rear legs, emphasizes this point. The other, larger male seems to have been caught off guard by the contender's move, as he rounds in front of the attacker, lifting his head, tusks, and trunk in what Akeley seems to have imagined as a threatening stance. Again, while we would not see this position as signaling a threat today, Akeley's vision of two male elephants fighting should be understood through an aesthetic and conceptual framework rather than that of natural history. Akeley portrays the sheer power and size of adult male African elephants—animals much larger than the females we are accustomed to seeing in zoos. Raised off the floor, and with the larger elephant's trunk reaching spectacularly high into the Rotunda space for which the work was designed, the piece was arguably more suited to the old Columbian Fine Arts Building than to the eventual Field Museum, built ten years after it was created, where the work still stands, now next to the even more spectacular *Tyrannosaurus,* Sue.

29 Carl Akeley, *The Wounded Comrade,* 1913. Bronze sculpture, 24 x 12 ¼ in. The Field Museum, Chicago.

As he recalls in his memoir, *In Brightest Africa,* from early on Akeley had hoped to "start taxidermy on the road from a trade to an art," and clearly *The Fighting Bulls* was part of that effort. After he left the Field Museum and began work at the AMNH, however, he claims that he became aware that "the kind of men who supported great museum ventures" tended to support art museums more than taxidermy. If he wanted to realize his dream of a great African hall for animals, he felt, then potential benefactors would need to see his work as art. In the end, in an effort to "make an appeal to those men who support art financially," Akeley decided to create a small work in bronze. Recalling a story he had heard from a Major Harrison, he sculpted two elephants supporting a third, wounded elephant between them (fig. 29). According to Akeley, Harrison "was shooting in the Congo and came upon four big bulls. One he killed and another he wounded. The wounded one went down but the two survivors helped him regain his feet, and with one on each side helping him the three moved off." Akeley explained that the story appealed to him because it showed "a spirit in the elephant that I should like to record. I set to work on The Wounded Comrade. It was a part of the story of the elephant, a theme that always aroused enthusiasm for me."[13] The work shows three elephants moving slowly forward, the outer two animals pressing up against the largest of the three, the wounded elephant in the center. The face of the elephant on the right is obscured behind the ear of the central figure, as every muscle in his body, and even the earth itself, appears to press in to support the weakened giant. The face of the large bull in the middle has a sunken, resigned look, his trunk hangs listlessly, and his limbs seem thin and shaky. Finally, the elephant on the left has placed his right tusk under the trunk of the wounded comrade, while his trunk wraps around and holds the central elephant close. Akeley argues that he was able to portray subjects in bronze that, because of the immense amount of time required, he could never create through taxidermy. But, while working in bronze was efficient, Akeley clearly understood the two types of work, as Alvey has shown, as complementary if not essentially the same. In both media he was able to show not simply the physical form of the elephant but also something both less tangible and more profound. Significantly, as with *The Fighting Bulls,* his first bronze sculpture depicted something about the creatures that he himself had not observed. Far from being the fastidious describer of nature, then, in these—some of his most famous works—Akeley wanted to tell "the story." That story turned centrally on hunting.

In the spring of 1914 a controversy began to brew in New York City surrounding an eighteen-year-old Asian elephant named Gunda. Gunda had come to the Bronx Zoo in 1904, where he apparently settled in and for several years served as a ride elephant for children. As he matured, and not atypically for male elephants, Gunda became increasingly difficult to manage. He attacked the head keeper twice, almost killing him in both cases, and as his periods of musth grew longer and more intense, the Bronx Zoo director, William Temple Hornaday, decided to keep Gunda closely chained to a wall. Not surprisingly, visitors to the zoo began to complain about Gunda's circumstances and letters began to flow to the *New York Times,* other newspapers, and the SPCA. In a June 1914 article titled "Bronx Zoo Elephant Chained for 2 Years," the *Times* reflected that since elephants can live for eighty years, Gunda had little more to look forward to than "standing in the Bronx Zoo with one foreleg and one hindleg chained to the floor for the next sixty-two years." The article continued, "Many of the persons who flock to his cage and watch him as he tests the chains that bind him stand fascinated by the silent drama. He is the biggest elephant in captivity on this continent. He has more strength of bone and sinew than any other living creature. Yet he can move only two feet forward and two feet back, and then he drops his massive head and heaves from side to side, at once a grotesque and tragic figure."[14] A letter to the editor three days later claimed, "It is impossible for any human being with ordinary common sense not to see plainly the misery of this intelligent and powerful beast as it stands 'forever chained to one damned spot.' That this is an intolerable cruelty is unquestionably true." For this writer, the choice was clear: "If Gunda is incurably vicious, as it is asserted, then in the name of decency shoot him, and make of him a magnificent 'stuffed' specimen. . . . The public's 'education' can be accomplished quite as well by the sight of Gunda dead as alive, and we shall be spared the sight of a needless suffering that is as unpardonable as it is degrading."[15]

Hornaday came back swinging. In a letter that appeared the next day he wrote mockingly, "Won't it be a spectacle for gods and men when Gunda, munching hay in the most comfortable animal room in the whole Zoological Park, hears the roar of the elephant rifle that sends a steel bullet crashing through his skull and brain in the name of The Merciful, The Compassionate? When it is done, I shall insist that those who are now attacking my management of Gunda shall be compelled to come to the death chamber, and see it done. As for myself, I will ask to be excused. I think Gunda's days are numbered; and when he is finally shot to death we will erect a tablet to his memory bearing the inscription: 'Shot to death in response to the demands of tender-hearted friends who wanted him to be free to turn completely around in his room.'"[16] For Hornaday, the only two reasonable choices for Gunda were to keep him chained in the hope that he would return to a more tractable state or to shoot him where he stood. Never keen about taking advice from "sentimentalists," Hornaday seems nevertheless to have believed that Gunda's musth could be managed and that the animal would remain a valuable part of the collection.

A year later, though, Hornaday relented. Gunda had long since ceased to be the focus of editorial acrimony; he was yesterday's news. On 23 June 1915 the *Times* ran a story under the headline "Bullet Ends Gunda, Bronx Zoo Elephant," which began, "The troubles of Gunda, the bad elephant of the Zoological Park in the Bronx, ended yesterday morning, when from the vantage point of the small iron door leading from the side of the elephant house into his inclosure, Carl E. Akeley, Assistant Curator of the American Museum of Natural History, sent a bullet crashing into Gunda's brain." The article reported that Hornaday had become convinced that Gunda was suffering, that he was refusing food, and that "it was apparent that he was no longer any use as a zoological attraction." At 8:00 A.M., after the other elephants had been moved outside, Akeley stepped through a small door into the exhibit. Gunda stood, chained in place. According to the *Times,* Akeley

> used an elephant rifle, a Mannlicher, 25.6. Gunda turned his head as the small door opened and glared at Mr. Akeley for a second, and then, straining at his chains, he made a vicious swing with his trunk at his executioner. Mr. Akeley is a distinguished elephant hunter, and

> as Gunda lifted his trunk he raised his rifle and fired. The steel bullet caught the elephant between the eye and the ear, penetrating his brain. The uplifted trunk stopped in the middle of a sweep, the animal closed his wicked little eyes, opened them and then, quivering, sank down and died practically without a struggle. Director Hornaday and Keeper Walter Thuman went out while the animal was being killed.[17]

The article quoted Hornaday, who, revealingly, claimed that Gunda "was painlessly killed. The shot produced instantaneous paralysis of the brain. The huge beast simply dropped in his chains and died without a struggle." What stands out in the *Times* reporting of the execution is Akeley's swiftness and skill and the intensely slow-motion death of the elephant. After describing Gunda's closing and opening his eyes, and the quiver that went through his body as he sank and died "practically" without a struggle, Hornaday, who was not at the scene, is seen managing public reactions by emphasizing the "instantaneous paralysis" of the animal's brain.

For many people, a figure like Carl Akeley seems paradoxical. Here is a man who takes a photograph of his wife as the calm hunter standing on her first elephant; who has pictures taken of himself lounging, looking amused, on the body of the same animal; who walks into a small room at the Bronx Zoo and executes a chained elephant with a single shot to its head while the animal's keeper and the zoo director feel they have to absent themselves; and who creates a work of art like *The Wounded Comrade,* which asks us to question the figure not represented: the hunter, the persecutor, the killer, the human. Alvey notes the irony that Akeley believed that in order to preserve the animals of Africa he had to kill them.[18] The irony extends, though, to a broader issue in trophy shooting and trophy shots. However much these activities and the works about them seem to function in a fairly limited emotional register—triumph, elation, satisfaction, desire—there has always been something resting behind the hunts and their images that poses questions, and those questions inevitably play out in the public display of hunting.

This is why Hornaday felt it necessary to respond to rumors about Theodore Roosevelt's hunting activities in Africa. In a 1913 letter to the editor of the *Times,* he insisted that Roosevelt's hunting in Africa had nothing to do with some unwholesome desire to kill animals. Recalling his response to the news of Roosevelt's expedition, Hornaday wrote that he "rejoiced that at last a vigorous, red-blooded, hard-working, and thoroughly conscientious American Sportsman was willing to go to Africa and work like a slave for the pleasure of filling some of the empty cases of the National Museum with fine wild specimens."[19] That

Roosevelt worked "like a slave" is obviously nonsense, but Hornaday clearly felt that Roosevelt's reputation—his public standing as "vigorous, red-blooded, hard-working, and thoroughly conscientious"—was at stake in the rumors about the expedition, and he wanted to be on record defending the Colonel's honor. The reason for this conviction is that Roosevelt's hunting, like all trophy hunting, but certainly more so because of his fame, was a *public* spectacle, not a private hunt.

In December 2008, waiting for a long-delayed plane, I began a conversation with two middle-aged men who, it was quite evident, were hunters who regularly traveled to pursue large game. Our conversation touched on everything from black bears in the north woods of Wisconsin to African safaris. One of the men told me that his uncle was at the time preparing to travel to Africa for the purpose of shooting a bull elephant. Surprised, I asked how much it might cost to go on such a hunt. I was told that, excluding airfare, the cost would end up in the range of $30,000. While acknowledging that that was a lot of money, he averred that there aren't too many people who can say they've killed an elephant. The critical point of trophy hunting, as opposed to hunting for the table or even for scientific purposes, is that however much the trophy hunter might like to imagine and portray himself as alone in the bush (as if the trackers, translators, gun bearers, servants, cooks, and these days even videographers were not also there), these hunts are fundamentally about the performance of a spectacle *back home.* Sometimes, as in the case of Roosevelt, the audience is huge; sometimes, as in the case of a woman who became the first person to kill an elephant with a bow and arrow, the audience might grow well beyond expectation and desire; but usually the audience consists of "hunting buddies," family, and friends.

Indeed, the public nature of these activities is as apparent in the famous hunting narratives of the late nineteenth century as it is in hunting websites, magazines, and Flickr accounts of the twenty-first, and it is precisely because of public scrutiny that hunters are so preoccupied with both the elaborate rituals surrounding hunting and the carefully articulated (if constantly shifting) rules of sportsmanship involved. The key here is that rituals and rules stem not from hunters staring at the game with clear-eyed enthusiasm but from *doubt,* a feeling that is echoed, I believe, in one of the more remarkable photographs from Akeley's first expedition to Africa, in 1896, for the Field.[20] The photograph, reproduced in Alvey's article, shows a dead Somali wild ass lying on its left side on the ground (fig. 30). The striping of the animal's forelegs, part of the critical fascination of the species for museums in the period, can be seen at the left of the image, while the animal's rear end and hind legs fall outside the frame. The ass's eyes are open, its ears are back, and the tensed neck suggests an agonal struggle, as does the animal's

30 *Somali Wild Ass,* 1896. Photograph. The Field Museum, Chicago.

lower lip, which has bunched up in the dirt. As much as the work is a photograph of the dead ass, it is also a photograph, as Alvey points out, of Akeley, whose shadow is prominent below the arch of the animal's neck and head.

It seems clear from Akeley's memoir that he wanted explicitly to implicate himself in this particular death. In general, Akeley did not see himself as a sport hunter, except perhaps when it came to elephants and lions, because in those hunts, he argued, he "always felt that the animal had a sufficient chance in the game to make it something like a sporting proposition." When it came to other kinds of shooting, however, he claimed that it often felt "a great deal like a murder." To emphasize this point, he recalled two experiences collecting Somali wild asses that left him feeling that he "had had quite enough" and that "if this was sport, I should never become a sportsman." Describing the shooting of the second animal, the one pictured in the photograph, Akeley writes, "Just at dusk the shadowy forms of five asses dashed across our path fifty yards away and we heard a bullet strike as we took a snap at them. The one soon stopped and we approached keeping him covered in case he attempted to bolt. As we got near he turned and faced us with great, gentle eyes. Without the least sign of fear or anger he seemed to wonder why we had harmed him."[21]

One response to writing like this is to claim that Akeley was simply attempting to excuse what he thought others might perceive as unsportsmanlike behavior. That is, he didn't really care about the animal, but he thought others might. Another way to think about the photograph, about Akeley's writing, about his taxidermic work, and about the photograph of Delia, is to accept that these works are always about people as much as, and sometimes a great deal more than, they are about animals. Critics who argue that trophy shots demonstrate the total objectification of animals in necrophilic hunting miss the essential point of these images. The images are attempts to make sense of the hunt for both the hunter and his or her audience at home. It is not that the animals don't matter in these works—the animals and their deaths are absolutely the central moral, emotional, and intellectual *matter.* But these works have never been about simply displaying death; they are attempts to make sense of death. The sense they make—constellations of ideas both related to and different from other portrayals of animal death, including a work like Douglas Gordon's *Play Dead; Real Time*—is always specific to a time, a place, an animal, a hunter, a collector, a gallery visitor. It is this specificity that must be at the core of our efforts to understand the profound ways in which people engage the world around them through their interactions with and ideas about animals.

Fishing for Biomass

SAJAY SAMUEL AND DEAN BAVINGTON

Chapter EIGHT

The Grand Banks off the island of Newfoundland on Canada's east coast once teemed with such an abundance of codfish that they reportedly choked the passage of vessels. About five hundred years after the discovery of the New World, in 1992, fishing trawlers could find no cod, and officials declared a moratorium on the fishery. Little has changed since the disappearance of wild cod from the seas off Newfoundland and Labrador: both the fish and those who relied on hunting them are hard to find.

From the mid-nineteenth century on, the cod fisheries on the Grand Banks have been the object of techno-scientific management. Processed through such algebraic machines as merchant capital, jiggers, and scientific models, codfish have been transmuted into biomass, fishermen transmogrified into labor, and coastal waters transformed into liquid farmlands. Merchant capital demanded that fish be harvested in the predictable patterns and quantity necessary to meet the schedule of loan repayment. The resulting outsized technologies increased catch levels enough not only to pay off financiers occasionally but also to fatten profits. International agreements established national boundaries to police fishermen and capitalists, both foreign and domestic. And scientific theories of fish and men precisely calculated the maximum quantities of fish that could be harvested from the oceans, in perpetuity.

Yet, within three generations of scientific management, the Grand Banks were emptied of commercially viable stocks of codfish, and they have remained empty for two decades. In the absence of cod, the laborers once employed on trawlers and in fish factories have been redistributed as oil workers to the tar sands of Alberta, professional harvesters to the aquaculture industry, and tourist guides to the oceans. And those who still want to fish cod for food are criminalized.[1] Without grasping the crucible in which codfish are transmogrified into saleable biomass, these fishermen have little to hope for.

A World Without Fish

Worldwide, the oceans are emptying of fish. Since the 1990s the catch of wild fish has declined steadily. Nowadays, even UN bureaucrats and fisheries scientists confirm with statistics what fishermen have been warning for years: that the waters of the earth are being emptied of their creatures. Some 77 percent of global fish stocks are fully exploited or have collapsed.[2] Authoritative opinion suggests that all commercially valuable species will collapse within a generation if current trends continue.[3] Some four million commercial fishing vessels trawl the oceans, capturing some 350 billion pounds of fish annually, 25 percent of which is discarded as waste, quaintly called "bycatch."[4] Predatory fish such as salmon, cod, and shark are being killed at rates that threaten their existence.[5] Those that remain are emaciated by hunger because the prey they feed on is being harvested as fishmeal to fatten chicken, cows, and other fish.[6] The growing consumption of omega-3 fats in the North is devastating the primary protein source of millions of fishing communities in the South.[7] Fully half of the fishy biomass served on tables in the North is now manufactured on watery farms.[8]

The collapse of the Newfoundland and Labrador cod fishery (once the world's largest ground fishery) has become legendary and reflects these global trends. The 1992 moratorium on cod fishing in these waters continues to this day. For almost two decades now, the cod have not returned to the Grand Banks and have recently been recommended as a candidate for Canada's endangered species list.[9] The disappearance of codfish from these waters was neither a natural disaster nor an accident. Hunting for sport decimated the bison that once roamed the North American plains in the millions, and the passenger pigeons that once blotted out the noonday sun during their migratory journeys. But wanton destruction does not account for the disappearance of the codfish. That was the result of the scientific management of a natural resource.

Yet the scientific management of codfish continues to be peddled as the cure for its destruction. The failure *of* scientific resource management has been reinterpreted as a failure *in* it, prompting the search for new and improved versions of resource management. Accordingly, by overwhelming scientific consensus, "overfishing" is now blamed for the disappearance of the cod. Rapacious industrialists, greedy consumers, ignorant fishermen, and pliant government bureaucrats have been fingered as the usual suspects in this familiar narrative, which takes as given objective science and neutral technology.[10] By constructing the problem as a matter of "overfishing," more refined scientific management of cod fishing becomes a plausible solution. Conveniently forgotten in this familiar story of capital's misuse of science and technology is the fact that until 1992, the cod fisheries off the coast of Newfoundland and Labrador were generally regarded as outstanding examples of the scientific management of a natural resource.

Thus the contemporary social imaginary runs on a narrow gauge where techno-science constitutes both rails of problem and solution.[11] For instance, the alternative to the permanent decline in wild cod stocks is to produce cod in egg-to-plate production systems that begin on fish farms. Such farmed cod are less like lions in a zoo used to titillate city folk than like domesticated game reared and shot in pens by trophy hunters. This aquaculture of cod is only the most recent phase in a longer history of the scientific management of cod as natural resource.

Killing Cod Scientifically

For most of human history, fisheries have been neither managed nor objects of techno-science. To this day, fishing is primarily a small-scale subsistence activity, subject to the rhythms of season and tide, subservient to the nature of fish, and limited by the ability of fishermen. By the latest count, such fishermen outnumber laborers on the world's industrialized fishing fleets by a factor of sixty. They receive only one-fifth of the public money given to the industrial fleets, yet they catch as much fish. They produce none of the pollution routinely generated by industrial fishing, avoid harm to spawning fish and habitat, and waste none of their catch. Oriented primarily toward fishing for food instead of for profit, the wide array of artisanal fisheries, which include the stilt fishermen of Sri Lanka, the horseback shrimpers of Belgium, the poisoned-spear fishing of the now extinct Marsh Arabs of Iraq, and cormorant fishing in China, do not mistake fish for a commodity, fishermen for workers, or fishing for employment.[12]

Similarly, from the late eighteenth century until World War II, cod fishing off Newfoundland and Labrador was practiced predominantly by subsistence-oriented fishing families who produced salt-dried cod for trade with merchants.[13] These fishing grounds were once the richest and most productive in the world, intersected by the north-flowing warm Gulf Stream and the south-flowing cooler Labrador Current. Rich in microorganisms and minerals, these waters once supported such an abundance of marine life that only "in Newfoundland . . . can one be made to feel the contrast between a land that is infinitely silent, motionless, poor in vegetation, above all poor in its variety of living creatures, and a sea which harbors every form of life."[14]

Unlike herring and salmon, the cod has a high protein content without being fatty, for which reason it was once called "the beef of the sea." As a lean fish, it is also relatively easier to salt-cure and dry in the sun, which accounts for the centrality of cod to the Atlantic fish trade between North America, Spain, and Portugal, among other countries, in the early modern period. No wonder, then, that for half a millennium salted and dried cod, variously called *bacalao* (Spanish), *bacalhau* (Portuguese), *klippfisk* (Norwegian), and þorski (Icelandic), has constituted the staple food in many a cuisine.

Codfish are carnivores. As fry they feed on plankton near the surface of the water, graduating as they grow to eat small crustaceans, including mollusks, and then to bigger fish such as herring, capelin, and squid. Codfish favor saltwater and thrive in temperatures around fifty degrees Celsius, though warmer waters speed up their hatching and reproductive cycles. Location, climatic conditions, and the salinity and warmth of the waters largely determine the spawning, growing, and reproductive cycles of codfish. Cod in the Gulf of Maine spawn in winter, while those in the Gulf of St. Lawrence spawn in summer; the shore cod of Newfoundland reach adulthood in about two years, whereas bank cod take about four years to mature. Though codfish average about two feet and some six pounds in the Grand Banks, they vary in length and weight; one of the largest cod caught off the Labrador coast was recorded at more than five feet long and above one hundred pounds.[15]

Governed by their breeding and feeding cycles, hunting for cod was circumscribed by the times and seasons during which codfish were available. Typically, the inshore fishing season was limited to the roughly six to eight weeks when "the cod-fish bound[ed] above the waves, and reflect[ed] the light of the moon from their silvery surface" as they chased in from the banks their main food source, the capelin that were "hurrying away in immense shoals to seek refuge on the shore."[16] Fishing relied on a baited hook on a single line dropped over the side of a small rowboat called a dory. Though larger boats were used seasonally to reach the offshore fishing banks, the technique of hunting cod with baited hooks necessarily meant that the cod had to be hungry enough to go for the bait. Once cod gorged themselves on capelin, they would stop biting at the baited hooks and the cod-fishing season would come to an end. Well into the twentieth century, when the fish were out at sea, fishermen would repair boats and houses and find other ways to subsist, including working garden farms of potatoes and turnips, raising livestock such as pigs, goats, and chicken, and occasionally pursuing caribou during the winter.[17]

Algebraic Machines I: Finance Capital

By the middle of the eighteenth century, the cod fisheries off the Grand Banks were tightly woven into trade routes connecting Europe, Africa, North America, and the Caribbean.[18] Fish, rum, slaves, and manufactured goods were exchanged according to variable equations, and merchant capitalists from England and other European countries financed many of the Newfoundland cod fisheries. For most of nineteenth century and into the twentieth, the political economy of the cod fisheries is best characterized as merchant capitalism. Through the store, the merchant sold fishermen such necessaries as fishing supplies, building materials, and foodstuffs, including tea, flour, salt, pork, and lamp oil. However, they did so mostly on credit instead of for cash. The sale of goods from merchant to fisherman, from roughly the beginning of spring through the fishing season, raised a legally enforceable claim against the total catch of the fisherman in the fall. Called the "truck" system, this circulation of goods for codfish meant that merchants assigned prices for both sides of the trade.[19] The custom of borrowing money in the form of goods and repaying such credit in codfish meant that the logic of capital and interest was inserted into the fiber of the relation between fisherman and merchant. Although fishing families controlled the modes of producing salt cod, the value of goods received and cod given in payment was recorded and calibrated in the merchants' books of accounts. Should the value of cod exceed that of the goods given on credit, the merchant would balance the books by providing a "winter supply" of goods. Should the fisherman have received more goods than he paid in cod, he was carried on the books to the next year.[20]

Crucially, however, both the length of the fishing season and the quantity of codfish caught in a given season naturally fluctuated from year to year. This meant inevitable mismatches between the predictability of loan-repayment schedules and the variability in the comings and goings of the codfish. For many reasons, not least the increasing scale of the merchant's operations, the mismatch became increasingly unprofitable for the merchant capitalist, who demanded a more reliable and increased flow of codfish to match the growing amount of credit outstanding. The customary practice of carrying forward loans or forgiving debts because of unexpected changes in catch volumes reflected the subordination of finance capital to the exigencies of subsistence fishing and the vagaries of nature. The dream of disciplining the natural ebb and flow of codfish to the time value of credit money therefore presupposed a reversal in this hierarchy of values. By making loan repayments a normal and normative demand, finance capital began to reshape the forms-of-life[21] lived in concert with codfish.

Algebraic Machines II: Jiggers and More

In the mid-nineteenth century, fishermen could do little to change the rhythms of feeding, spawning, and the migration of codfish, attuned as these were to season, temperature, and other imponderables. However, they were tempted and encouraged to pursue the solution of increasing the catch. Specifically, in response to the demand to increase the annual catch, fishermen began to experiment with and adopt more intensive technologies for fishing. Thus, after close to 350 years using the baited hook-and-line method, the introduction of a slew of new instruments in the latter half of the nineteenth century changed the practice of fishing. All of the new technologies shared one crucial feature: they were designed to ignore the nature of codfish.

The efforts to increase the catch through technical means relied on both extending the fishing season in time and across space and increasing the quantity of the catch per unit effort. The cod jigger (two hooks attached to a lead weight emblazoned with the *image* of a capelin) and the cod seine were the first in this class of instruments. By design, the jigger was dropped into a school of satiated cod and rapidly moved up and down until one of the hooks pierced the flesh of a codfish so that it could be hauled aboard. With the jigger, cod could be captured after they were glutted with capelin, lying lethargically on the bottom of the fishing grounds and refusing to bite at baited hooks. The cod seine, a roughly six-hundred-foot-long net composed of four- to five-inch meshes, was towed by a crew of about six people, thus vastly increasing the productivity of labor. Entire schools of codfish could be hauled aboard with the seine, catching young and old, spawning and mature cod alike.

The jigger and the seine epitomize all industrialized fishing technologies aimed at increasing the catch size: the disregard of the nature of codfish and the disdain for the agonistic craft of hunting fish. Designed to catch fish whether they were hungry or not, the jigger and seine transformed codfish into biomass. Reciprocally, such technologies also entailed transforming the activity of fishing from hunting to harvesting, and recasting fishermen as abstract labor. A related fishing tool was the longline or bultow, which comprised a central longline attached to dozens of fishing lines and hooks dropped to various depths and kept afloat with glass buoys on either end. Whether the jigger, the longline, or traps, seines, and gill nets, the new technologies introduced during the second half of the nineteenth century transformed the activity of hunting codfish into the passive harvesting of a natural resource conceived as fishy biomass.

Yet it was only with the deployment of the first bottom dragger in 1954 that natural thresholds associated with the cod fishery were completely transcended. Although longlines today run hundreds of miles and have thousands of hooks at depths exceeding two thousand feet, it is the factory freezer dragger, trawling the ocean beds without regard for weather conditions, the hunger of cod, or other natural phenomena, that allows cod fishing to be temporally and spatially unbounded. The British built the first trawler or dragger in 1954—the *Fairtry*—and that design was soon copied and elaborated by the Russians, the French, the Japanese, and the Spanish. The *Fairtry* was more than 280 feet in length, lay at twenty-six hundred pounds of gross tonnage, and had onboard a freezer and filleting and fishmeal machines. All the processes, from harvesting to making the codfish market-ready, could now be conducted onboard, and the *Fairtry* was able to fill her hold with six hundred tons of fish in forty days.[22] The steady intensification in the industrialization of fishing technologies has had the desired consequence of increasing the catch. Of the hundred million tons of cod that is estimated to have been captured from 1500 to 1992, it took four hundred years to hunt down half of that amount, and less than a century to harvest the remaining fifty million tons.

Since the collapse of the cod in 1992, the idea of aquaculturing cod has taken over the imagination of fisheries scientists, investors, and government agencies. Aquaculture entails not only programming the production of biomass but also individual or corporate ownership of it.[23] In contrast to the older forms of "capture fishery," this new "culture fishery" on fish farms aims to control the complete life cycle of codfish from "egg to plate." Having failed to predict or control the behavior of codfish in the wild, the fish farm now imposes the controlled conditions of a scientific laboratory on the waters. Fish farms are thought to solve the problem of wild cod by reengineering them as a domesticated species.

Algebraic Machines III: Modeling Living Fish

Fisheries science runs in tandem and is intertwined with the intensification of fishing technologies. For most of the history of the cod fishery, knowledge of the fish was limited to what fishermen knew. Interannual fluctuations in landings were therefore accepted as the inevitable consequence of variable migration patterns, thought typical of all marine fish. Yet, by the 1850s, pressured by fish merchants to increase the catch size and by investors to smooth out fluctuations in the catch, governments began to recruit scientists to discover what caused interannual fluctuations in landings and to uncover the supposed "natural laws" that determined years of lean and plenty.

In the 1880s the German biologist Friedrich Heincke applied the statistical methods he borrowed from the study of human populations to observations of morphological features of herring that had been caught in different fishing locales. The first to self-consciously apply scientific methods to the study of fish, Heincke insisted that population thinking could be fruitfully adapted to the fisheries.[24] Codfish could be thought of as a single species population whose measured birth and death rates would produce statistical estimates of harvestable cod. The notion of "population" had so infected the field of fisheries science that by the 1950s population models predicted the amount of surplus biomass (or live weight of fish) produced each year by a specific fish population, labeled maximum sustainable yield (MSY), which if extracted would maintain stable annual landings.[25]

The scientific conceit behind this prediction was that annual landings would remain stable for perpetuity if no more than the predicted surplus biomass were captured. While the surplus-production model allowed fisheries scientists to determine the maximum sustainable yield of fish populations, it did not account for such technological innovations as the factory freezer trawler. Reality overtook that scientific model when, for instance, at the peak of the Newfoundland cod fishery, trawlers from more than twenty nations jostled for position on the offshore banks, creating a "city of lights" on the fishing grounds. By 1968 this "cod rush" off Newfoundland resulted in a "killer spike"—the largest annual landings of cod ever recorded—and a precipitous decline in landings thereafter.[26]

By reframing this decline as a "tragedy of the commons," economists joined biologists to widen the scope of fisheries science. While the surplus-production model allowed fisheries scientists to determine the maximum sustainable yield from fish populations, there was no way to guarantee that fishermen would behave accordingly. Economists, claiming to understand and model human behavior, then proposed bio-economic models as the appropriate scientific response to the failure of the prior population model.[27] By including the demand side of the equation, economists claimed to be able to offer a more robust prediction of the equilibrium quantity of landings. They further proposed limiting access to fish populations by enclosing the oceans within defined national boundaries. Creating public property of about two hundred miles of ocean excluded foreign fleets and gave legal teeth to quota restrictions imposed on domestic fishermen.[28] Within these constraints, rational fishermen could now maximize their profits while minimizing the dangers of overfishing. "Total allowable catch" and "maximum economic yield" were the resulting new and improved scientific numbers that now fell out of the system of equations developed to optimize the joint behavior of codfish and the fishing industry.

Notably, from its inception in the population model, fisheries science was built on the assumption that living beings can be described using the logic of mathematics and experiments. Maximum sustainable yield and total allowable catch were the glue that bound together industrialists, regulators, fishermen, fishing technologies, and fish. By harnessing outsized technologies and profit-seeking capitalists to such numbers and giving the supervisory reins to government regulators and scientists, it was believed that no more than the surplus biomass of cod would be harvested. Thus, on the eve of their collapse, the cod fisheries off the Grand Banks were widely acknowledged to be the most scientifically advanced, best-funded fisheries management regime in the world.

And yet, in 1992, the cod fisheries off the Grand Banks collapsed, apparently forever. This triggered the largest single-day layoff in Canadian history, bringing to an end the many forms of life shaped by cod for nearly half a millennium. Induced by unemployment and relocation programs, the fishing families of Newfoundland were broken up and dispersed across Canada as far as the tar sands of Alberta in search of food and shelter. Others were encouraged to seek remarkably ineffective retraining programs. For instance, of the some forty thousand persons receiving such assistance between 1994 and 1996, only "732 people were placed in jobs outside the fishery, 1,492 retired from the fishery, and 3,900 got jobs in non-groundfish fish plants."[29]

In the wake of the collapse, the assumptions of bio-economic population models have come under attack. Increasingly, fisheries science has become informed by the presuppositions of complex systems theory, which theoretically falsify many of the equilibrium and averaging assumptions of bio-economic models. Complex systems such as traffic and stock markets operate in nonlinear ways, and wild cod are now assumed to behave similarly. However, since ecosystem models with accurate predictive ability have not yet become available (and some scientists argue that they are theoretically impossible given the uncertainty and complexity of perturbed marine systems), bio-economic models remain the dominant scientific way of understanding wild cod, though they are used with increased caution and ironic recognition of their unreliability.[30]

Reviving Fishy Biomass

Since 1992, all attempts to respond to the collapse of the cod fisheries have relied on applying further doses of techno-science to maintain the profits from the production and sale of codfish. The "money value" of time, "biomass" of fish, and "population" of cod are abstractions that continue to be produced in the crucible of algebraic machines. Once, finance capital provoked technical innovations in the cod fisheries by making plausible a correspondence between the calculated flows in the value of time and the volume and constancy of fish landings. So provoked, fishing technologists continually sought to increase the catch by increasing the scale of operations. Exemplified by the jigger and culminating in the factory freezer trawlers, technical advances in fishing entailed staging codfish as biomass, fishermen as instances of abstract labor power, and fishing as work. The potential for large-scale technologies to destroy what they set out to harvest also sparked the birth of fisheries science in the late nineteenth century. Increasingly refined scientific models of codfish and fisherman behavior took for granted what technique had made possible, and offered precise calculations of the harvestable biomass. Despite its evident failure, that once ruinous alliance between economics, technology, and science continues.

For instance, the disappearance of wild codfish has been interpreted as an occasion to profit from raising codfish on fish farms. In this effort to create a cod aquaculture industry, scientists and entrepreneurs—leveraging university research and enabled by policies from provincial and federal Canadian governments—are beginning to domesticate cod populations thoroughly over their entire lifecycle. Cod are grown in hatcheries and brood stocks are developed through a cod genome project that has identified genetic traits consistent with rapid growth and large sizes suitable to maximize profits. Investments in techno-scientific inputs increase the resistance of such fishy biomass to diseases and stress-induced behavior, including cannibalism, produced by being confined in sea cages. Feeding cycles, water temperatures, and the training of workers all aim at optimal control of the conditions for the production of profitable biomass. Thus fish farms represent the effort to replicate the controlled conditions of a scientific laboratory on moving waters.[31]

Yet the dream of total control is unattainable, though not for want of trying. For instance, the effects of the interaction between the domesticated biomass produced through the artificial selection of traits geared to market values and wild cod remains unclear, particularly since the coastal grow-out sites licensed to produce farmed cod occur precisely where the remaining wild cod are found.[32] Fed on wild fish ground up as fishmeal, a pound of farmed cod indirectly absorbs between three and five pounds of wild fish. Thus, instead of preserving wild fish stocks, the aquaculturing of cod in fact depletes them.[33] Moreover, the industrial fish farms of the North beggar the subsistence fishermen of the South, since fish meal is made of small pelagic fish like sardines, anchovies, and herring, which constitute the staple food of many fishing communities in the South. Thus, instead of feeding the world's population, aquaculturing the "beef of the sea" contributes to starving it.[34] In a final irony, farmed codfish is packaged as part of the ideological system of ecological management. Yet fish farms, as realizations of controlled laboratories, potentially damage the environment. For instance, the waste from fish farms poisons the waters for miles around, much like that from hog farms on the land. Similarly, the cod that escape from their confinement in sea cages can change the genetic composition of wild cod, making the latter less fit to survive in their natural environment.

Moreover, egg-to-plate farming is highly capitalized and industrialized, producing the well-known consequences of corporate concentration. For example, to compete with low-cost producers like China and other white-fish products such as tilapia, cod farmed off the Grand Banks is now branded as a high-end product, following the success of marketing salmon. But the corporate organization of the Canadian salmon-farming industry is instructive in this regard: the roughly 140 start-up operations that began to farm salmon in the 1980s are now reduced to a group of some eleven giant corporations characterized by global ownership, vertical integration of operations, and contract farming.[35] Unsurprisingly, the individual fish farmer will be left holding an increasingly small share of the overall sale value of cod, most of which is captured by corporate intermediaries.[36]

The Fisherman Knows Best

So why, despite their manifest failure, do the algebraic machines of capital and techno-science continue to be marshaled in the cod fisheries? In light of the history of cod fisheries, "feeding the hungry," "technological efficiency," "conservation," and "scientific knowledge" are no more than the shibboleths of managers, regulators, scientists, and capitalists. A recent newspaper article is exemplary in this regard. Jeffrey Simpson, a well-known columnist for the *Toronto Globe and Mail,* inveighs against "irate fishermen," "the Fish Food and Allied Workers Union," and the "Provincial Fisheries Minister" for relying on "only anecdotal reports from fishermen" instead of on the scientific data produced by the Department of Fisheries and Oceans and endorsed by the Committee on the Status of Endangered Wildlife in Canada. In scolding these "local deniers and denigrators of [fisheries] science" for "discrediting the recommendations of the scientists," he complains, "year after year, indeed decade after decade, we stare in the face of scientific data that say stocks are declining fast, or improving only very, very slowly from decimated levels, yet the same voices remain in angry denial that more serious conservation action is required."[37]

To forget that it was precisely the most advanced fisheries science that purported to prove the abundance of codfish stock in the very year that cod became commercially extinct in the Grand Banks, Simpson reworks the familiar Enlightenment trope of the superiority of scientific knowledge to practical know-how. To support his case against fishermen who claim that there are enough fish stocks for them to resume inshore fishing, he characteristically conflates inshore fishermen with labor on a factory freezer trawler, confuses fishing for cod with harvesting biomass, and confounds the quantity of cod needed for local trade and sustenance with the tons needed to process in China and sell at Walmart.

Indeed, fishing communities have been the sites of dissension, debate, and protest since the introduction of the jigger, the seine, and the longline in the 1840s.[38] Without elaborating the details, one can uncover the red thread that runs through the fabric of these objections. For more than a hundred years after the introduction of the jigger, fishermen protested against using intensive fishing technologies on the grounds that they disregard both the codfish and the activity of fishing as such.

Jiggers and seines increase output, but they do so by indiscriminately killing and wounding codfish. Critics of killing spawning or satiated cod apply the same argument a fortiori to later, more intensive fishing technologies. Longlines and trawlers only redouble the disregard for hungry codfish that jiggers and seines encode by design. These methods now scoop up hundreds of tons of biomass from the oceans each year, regardless of species, age, or breeding cycle. What is not commercially valuable is thrown back, dead, as bycatch. Unmanned "ghost nets" that have broken free from trawlers now roam the seas unceasingly, netting a fishy biomass—entangling also whales and dolphins—destined to rot.

The objection to the capitalization of the fishery is not only that expensive technologies necessarily introduce wide differences of income and wealth. As important to the concerns of fishermen and other critics is the transformation of fishing by these technologies: when required to operate capital-intensive technologies instead of being able to use human-scale tools, fishermen are transmogrified into abstract labor. In the view of state officials and capitalists, the fisherman today can become a coal digger tomorrow and an environmental engineer the next day. Fishing technologies that unmoor the fisherman from his tools thus transmute him into a variable instance of abstract labor power.

Similarly, objections to mathematical knowledge are rooted in the difference between such scientific constructs as "population," "biomass," and "maximum sustainable yield" and concepts born of common sense and practical reason.

Scientific constructs not only erase commonly perceived distinctions but also create imaginary worlds—worlds, for instance, in which the notion of "ecosystem services provided by the coastline" becomes thinkable. This kind of scientific know-how remains a species of nonsense to the fisherman whose knowledge remains tethered to his embodied craft.

At the heart of their protests lies fishermen's recognition that fishing as a form-of-life can only be sustained by remaining within the natural thresholds defined by the feeding and breeding habits of codfish and their own embodied capacities. In this view, any kind of economic or techno-scientific rationalism that violates natural thresholds also subverts the nature of codfish and the culture of fishing. This approach views the fisherman as one who enters into an agonistic relation with his prey, and the activity of fishing is bounded by the limits of his ability and the cod's feeding and breeding habits. These dual natural thresholds demarcate the boundaries within which fishing occurs as a form-of-life. Beyond these bounds lies an undifferentiated space in which both fisherman and codfish are transmogrified. The fish farm is now built in that netherworld in which generic biomass is farmed and harvested as cod, abstract labor seeking utility masquerades as fishermen, and fishing can mean operating a six-thousand-ton factory freezer trawler.[39]

Today, the fisherman who hunts cod is a criminal under the law.[40] Hunting for codfish is criminalized because the hunter kills cod for largely noncommercial purposes. As such, he is a witness to a fading form-of-life that is either unthinkable or sentimentalized within the ambit of techno-scientific and economic rationalism. In that sphere, fishing with a baited hook and line is considered naïvely romantic, cussedly Luddite, or the willful stubbornness of "deniers and denigrators" of science.

Holding Hands

We do not claim to have clean hands. The fisherman whose complaints gave voice to the codfish, the fisherman whose practice embodies a lived relation with the fish, is left mute and largely defenseless by the operations of such algebraic machines as finance capital and techno-science. The question of what is to be done still looms large, unanswered. In the spirit of ecosystem-based management, some have sought to extract the local knowledge of fishermen in order to construct a more robust science.[41] Others have tried to translate scientific precepts into commonsense precepts through the education and professionalization of the fisherman.[42] When fishermen are not exploited as a source of knowledge for science or fashioned as professional managers, they are criminalized.

We have tried in this chapter to raise a barrier against those algebraic machines that have reduced and continue to reduce fishermen's forms-of-life to a form of life. Our argument that finance capital and techno-science necessarily transmogrify cod into biomass, fishermen into abstract labor, and fishing into work aims to fortify a line of defense against the nexus of power/knowledge that devastates a way of life lived in concert with fish. Let us not forget that the vast majority of fishermen and fishing communities of the world are oriented to subsistence and local trade. Accordingly, in the spirit of holding hands with those with whom we have little in common but to whom we are nevertheless committed, our exploration tells fishermen nothing that they don't already know.

Daniel Spoerri's *Carnival of Animals*

CECILIA NOVERO

Chapter **NINE**

> Who as a young child has not dreamt of the Frog-King or has not been afraid of the werewolf or yet still of Dracula who could metamorphose into a bat? Even Little Red Riding Hood tells the story of a metamorphosis of animal into human and vice versa. And then there are all those sayings, "as strong as a lion," "as cunning as a fox," "as filthy as a pig," etc., up to the common belief that the dog and his owner resemble each other. . . . And if we also think of all the masks present in various cultures and different carnivals, or if we think of all those mythologies in which gods and demons take the form of animals either to protect us or threaten us, marry us or rape us, then we have touched only summarily on a very controversial issue: I mean the fact that the genetic code is universal, namely, it is shared by bacteria, animals, and humans . . . [and that] all living creatures originally stemmed from the same cells, mankind included. We have not known this fact for long, since 1953 to be exact, the year the structure of the DNA was decoded.
>
> —**SPOERRI, *ANEKDOTOMANIA*, 153**

With this comment on the play between animal tales, proverbs, and science in his autobiographical *Anekdotomania,* Romanian-born, Europe-based artist Daniel Spoerri (b. 1930) suggests the genesis of his probing juxtaposition of animal and human physiognomy in his installation *Carnival of Animals* (1995).[1]

After being partially exhibited in various locations, *Carnival* is now permanently on view in a former monastery that the artist acquired and turned into a museum, in 2010, in the picturesque Austrian village of Hadersdorf am Kamp. Like many of Spoerri's other works, *Carnival* is made up of a series of assemblages in which found objects are mixed with two-dimensional images to create a new three-dimensional composition.[2] Thus *Carnival* follows the artist's principal techniques. These are *piéger,* "to trap," and *détromper,* "to undeceive." The former has characterized Spoerri's work from the start. It is particularly important because in the act of trapping the found situations of objects, the rule of chance substitutes for the artist's expert hand. Let me clarify. In his early *Trap-Paintings* (1960s) Spoerri glued onto the "tables" on which he ate the remainders of meals, including dirty dishes, glasses, bread, and ashtrays, exactly as chance had placed them there. He then hung these trapped and trapping tables vertically on his walls as artworks. For Spoerri, the snare tables thus suddenly entered a new field of vision: from remainders of life they became works of art. Hence viewers are encouraged to see the effects of their own act of consumption hanging vertically in front of their eyes, not beneath their gaze. Through this reversal of the horizontal to the vertical axis in the field of vision, art becomes visible as the outcome of consumption in life.

To undeceive, or *détromper*—a gesture that returns in several assemblages—is the act Spoerri employed to unmask the painterly illusion found especially in naturalistic art. More broadly, the works Spoerri names *détrompe l'oeil* defamiliarize the gaze that humans cast on the world, both in life and through artistic representation.[3] One key example combines the above-mentioned techniques and functions as a reference point for *Carnival of Animals.* Relying on the rule of chance, on the one hand, as in the trap paintings, and on *détromper,* on the other, Spoerri's *Criminal Investigations* (1972–91) toy with the proposition that virtually any object can become a murder weapon. For this series he enlarged photos from police archives on canvases, which he then overlaid with the rummaged objects from his wanderings through flea markets.

Similarly, in the *Carnival* series of wall assemblages, Spoerri recycled Charles Le Brun's canonical seventeenth-century drawings of animal and human physiognomic resemblances. As he explains in *Anekdotomania,* he first found these drawings by chance in a volume of the Swiss German poet, physiognomist, and theologian Johann Caspar Lavater's *Physiognomic Fragments for Furthering the Knowledge and Love of Man* (1775–78), a work Spoerri had acquired in an antique store. As he then adds, the drawings had illustrated a now lost 1671 lecture on physiognomy by Le Brun, first painter to the French king, held in Paris at the recently founded Royal Academy of Painting and

Sculpture.[4] Finally, Spoerri explains, he used the individual plates from Le Brun as the ground of each assemblage of animal figures in the *Carnival* series.[5] On top of Le Brun's animal-human resemblances, Spoerri playfully attached animal *trouvailles* (found objects). Each of the randomly chosen objects—animal objects, folkloric masks, avant-garde mementos, and scientific instruments, with their numerous cultural and personal meanings—had previously been used and circulated. Among Spoerri's creations we find a wooden rococo pedestal with lion carvings surmounted by a skull, which accompanies the lion and man comparison; kitsch ceramic owls glued to the study of owls and owl-men; a papier-mâché pig's head cane with a toy pig handle and a skull with tusks overlaid on the pig-man; and negative and positive battery cables, flowing from the spouts of metal faucets and attached to the drawn heads of monkeys and monkey-men. Because they challenge hierarchies, Spoerri's animal assemblages aptly take the title of *Carnival,* a title that immediately brings to mind popular disorder and the festive subversion of (all modern) classification, subversions in which animal figures have always also played a role—as masks, costumes, avatars, totems, and caricatures. *Carnival* humorously plays with the classification and order of things—especially animals—established by a canonical regime of vision that also grounds knowledge in Western civilization.[6] Indeed, one could argue that *Carnival* classifies animals, too, but this "bizarre" classification—like folkloric traditions—in the end produces masks (and fetishes). In doing so, it displays how human scientific truths about animals are in fact as carnivalesque as popular beliefs, and as lethal to animals, if not more so. The series of assemblages may occasionally, and significantly, provoke laughter, especially if one considers the contrast that the various mundane objects produce against the "historical" and "important" background of Le Brun's plates; through such a light approach to the serious issue of animal and human relations, *Carnival* shows how human conventions about "seeing" animals trap, on the one hand, humans in prefigured and myopic forms of knowing animals, and, on the other, animals in fixed and deadly categories.

Both the principles and the techniques of trapping and *détromper* that guide the whole of Spoerri's art can then help to answer a central question posed by *Carnival:* what are the interrelations within these assemblages between the animals and animal-humans in the background and the random, found animal objects in the foreground? Or, more precisely, what does the superimposition of object-animals do to the historical—that is, recycled and metamorphosed—animal physiognomy? To explore these questions, we need to recall the artist's location within the community of contemporary avant-garde and neo-avant-garde artists, and especially their critique of vision, which

Spoerri pursues here by way of animal-human contrasts and similarities, deriving from the physiognomic tradition's logic of resemblance. Strategically, temporally, parodically, but always critically, Spoerri explores how animals have been made to look in human representations—in particular, where appearance is joined to substance and character to soul. For physiognomy—whether famously, as in Giambattista della Porta's sixteenth-century studies, Le Brun's seventeenth-century speculations, or Lavater's pseudoscientific eighteenth-century analysis—is nothing less than an assessment of a person's character and personality by reference to the outer appearance, especially the face, i.e., a visible surface. With *Carnival* Spoerri continues his long-standing critique of rational vision that in this series of assemblages he explicitly links with the illusionistic resemblances drawn by physiognomy. Indeed, Spoerri takes pains to comment on the fact that the animal-human resemblances in Le Brun's drawings (1) are devoted to just the muzzle/face of the represented subjects and thus do not include the body, and (2) are immediately mapped geometrically, and specifically with the aid of "triangles" that Le Brun drafted on the muzzles of animals. Spoerri explains the geometrical mapping of the muzzle as follows: "The human being is the quintessence of creation. And in the head of a human being everything comes to be: 'caput ipse homo' [the human being is the head]. This conviction apparently kept Le Brun from comparing animal and human bodies."[7] And, one may add, from using triangles for human heads, where he instead drew straight lines (more on this below). While animals have resisted one-eyed (Godlike, and rational human) vision from its inception, as Paula Young Lee puts it, processes of mapping the animal "face" and fixing it into classifiable expressions—generally volatile and hard to capture—were the stake of serious (academic) representational art (in Jennifer Montagu's words) until at least the eighteenth century. In *Carnival,* Spoerri's intrusions of three-dimensional animal objects and bodies (corpses, masks, totems, etc.) on Le Brun's geometrically calculated physiognomic drawings (and resemblances) show the instability of human, rational vision.[8]

Indeed, adopting and adapting the *détrompe l'oeil, Carnival* challenges the viewers' habits of sight, as the former paintings had done. In order to question the relation between realism and reality, so called, Spoerri created *détrompe l'oeil* by gluing onto banal realist paintings found at the flea market equally banal objects, tools, and even embalmed animals or animal skeletons. In one instance he juxtaposed a showerhead with hose and faucets onto a landscape depicting a mountain with a flowing river. The man-made shower points to the disjunction between the imaginary, idyllic landscape of nature and art and a practical modern world in which water falls from a showerhead rather than from the heavens and flows through faucets rather than rivers. Objects and images, foreground and background, contaminate one another—as Spoerri himself puts it.[9] Their meanings are suddenly a point of contention, a question or riddle emerging directly from what seems most banal and evident, that is, realism and reality. In the *détrompe l'oeil,* then, a constellation of objects and images constructs a context—in fact tells a story—of reciprocal contamination. In *Carnival,* the history of human and animal relations is similarly tainted by the privileged ways of seeing and thus knowing animals.

Carnival also deploys the avant-garde principle of "objective chance" (*l'hazard,* a principle Spoerri inherited from surrealism). Spoerri's assemblages abolish any hierarchical distinction between the sciences and art or the pursuit of "mathematical" truth, in one case, and the exploration of unbounded imagination, in the other. Likewise, in contrast to the theological image of the great chain of being, still operative in Le Brun's age, or the Darwinian concept of natural selection two centuries later, Spoerri's *Carnival* features an assemblage of nonhierarchically ordered elements, or of historically unnecessary "connections" among all living creatures but also nonliving entities (objects, fetishes, idols).[10] A later work makes this point explicit: *Catena genetica del mercato delle pulci* (*Genetic Chain of the Flea Market,* 2000) presents a sixty-two-and-a-half-meter-long and one-meter-high frieze of found and ready-made objects that Spoerri had accumulated over thirty-five years. The human genome is applied here to the equally crucial life of objects. Like *Carnival,* the assemblage of random objects is ironic, absurdist, and fun, but also inquisitive and critical.

Le Brun's Drawings: Mapping Animals

As Montagu points out, Le Brun's drawings of animals and animal resemblances are indebted to the scientific/philosophical paradigms of the time, while at the same time they presented important innovations for the study of expression in art in an age when naturalistic representation was the aim of high art.[11] Three key elements arise here with regard to the function Le Brun has in Spoerri: (1) the relevance of Descartes for Le Brun, and Le Brun's fondness for classification, (2) the deep structural connectedness between art and science, both united in pinpointing and rehearsing the distinction between humans and animals, and (3) the didactic objective of Le Brun's drawings in matters of "expression." Spoerri's *Carnival* tackles these three points at the origins of modern thought, which he both ridicules and shows as "lethal" to animals. The impact of Descartes on Le Brun is especially important in the case of *Carnival.*

Descartes had located the pineal gland as the seat of the soul at the center of the brain. Further, the soul controlled the reactions of the body through the movements of the pineal gland. "If the passions were controlled from the brain, then the face, being the nearest part of the body to the brain, should be the most accurate index of the mind, and all of the soul."[12] The movements of the eyes and mouth toward or away from the seat of the soul in the pineal gland guided Le Brun's drawings, as his "diagrammatic [human] heads" show. These heads—traversed by straight lines—are transformed into geometrical maps when Le Brun portrays animals. In order to determine the character of each species, Le Brun indeed devised a system of triangles, which he accordingly used to draw the animals' heads. Triangles drawn with lines connecting various parts of the face—the organs of smell and sight and hearing—would visualize degrees of power and intelligence in animals. Le Brun dissected animal heads to demonstrate scientifically the veracity of his geometrical mapping, as far as the animals were concerned. As Montagu puts it, "Animals depend principally on the sense of smell, and secondly on that of sight. . . . Le Brun was . . . aware that this was not so of man." Hence he compared his animal heads with the sixteenth-century physician Andreas Vesalius's anatomical studies of the human head. In these studies Vesalius had outlined the pineal gland in red. Montagu adds, "For this reason Le Brun drew the triangle on the human head as based on a line running through the eyes, which precludes the possibility of drawing the other lines which incorporate the ears, noses and mouths into the system, and which revealed the characters of the beasts." The triangles Le Brun drew only on the animals' heads were thus aimed to render accurately the differences between humans and animals, even when the passions were the same. While the connections between the pineal gland and the eyes through the longer optical nerve in humans—as Vesalius's studies indicated—generated straight lines across the human eyes, the lines connecting the sensory organs of animals produced triangles.[13] The line through the human eyes pointed to the direct connection of vision with the brain, thus enforcing the belief in vision as the ontological ground of human being. Hence the brain and the eyes almost come to constitute a perspectival field particular to the human; its depth is indicated by the extended length of the optical nerve itself, hidden from view; for their part, the animals' heads (and animal expressions) appear to be all foreground, nothing but a face the depth of which is absent.[14]

The animal is the pure exteriority—or visible mask—of the more complex interior mechanism that generates the human, equally master of vision and emotions. Externally, animal and human face may mirror each other (albeit always through deformation, as per the eighteenth-century Swedish botanist

Carl Linnaeus). Ultimately, however, animals lack the depth that makes humans—and human vision—"not animal." Both interior and exterior seem necessary to open the vista on human-animal relations (resemblances), with the human taking the position of the fundamental "ground"—the ontological foundation of being, and vision; the animal instead is the mirror or mask of that core, the figure in the foreground, lacking foundation. Deploying devices that decenter the human, Spoerri's *Carnival* complicates the physiognomic perspectival game of mirroring at work in Le Brun's drawings: the transparency of this mirroring is here obfuscated, its violence exposed if not escaped. Like other *détrompe l'oeil, Carnival* performs vision as itself a *mise-en-abîme* of vision's own mechanics. *Carnival* multiplies resemblances. It does so first by proliferating animal objects, representations, and their functions throughout the series. Second, animals recur through their constitutive physical absence: the animal objects in the series are supplements that point up the erasure of living animals. The field of vision that *Carnival* opens appears to be already saturated by cultural icons and totems. It is replete with the ethnographic, scientific, and artistic mappings of the animal. These never-ending representations then interfere with Le Brun's resemblances, distorting and quite literally mortifying, i.e., tearing asunder their visible—recognizable, obvious—mathematical relations. As Spoerri's foxes exemplify, this infinite redoubling shows the laughable absurdity—the emptiness—of the human "ground" and knowledge, namely, the human wish to "capture" and "grasp" the animal. In this laughable absurdity, *Carnival*'s animals find their line of flight, as Gilles Deleuze and Félix Guattari would have it.[15]

Triangulating the Fox

In the fox assemblage (fig. 31), a dead specimen hangs from a common hanger, below a lady's hat: a middle-class woman's trophy of elegance, its eyes dead yet piercing, staring straight into the beholder's. Another fox is glued horizontally across Le Brun's drawings. The corpse, exposing its triangular jawbones, occupies the intermediary space between Le Brun's animal's visage and the fox-man's face. Le Brun's fox-man is the product of the resemblances between foxes (cunning) and clever men. Multiplied fake eyes and jawbones are scattered across the support, as unused tools, jaws defanged and in the geometrical shape that Le Brun adopted to map the animal's passions. The fox's tamed, i.e., disembodied, teeth are, in this assemblage, an accusing witness to an unfair struggle.

31 Daniel Spoerri, *Carnival of Animals: Reproductions of Humans Compared with Those of Foxes,* 1995. Assemblage on enlarged scan, after Charles Le Brun (1670), 185 x 100 x 30 cm.

As already mentioned, in other *détrompe l'oeil* (e.g., Spoerri's *Criminal Investigations,* 1972–91, or his *Anatomical Cabinet,* 1985–2001), the objects juxtaposed to the images' ground either break the ideal landscape represented there by debasing it or, alternatively, transform themselves into weapons or surgical instruments that provide the depicted body with an all too real substance: the body as injured.[16] Similarly, in the fox assemblage, the jawbones transform the geometrical triangles of Le Brun's rational calculations into a *corpus delicti.* A tool for calculating resemblances and for performing unequal exchanges between humans and animals, the triangle returns as a weapon that, having killed the animal, changes the fox into yet another, i.e., social, supplement: into the commodity fetish of a bourgeois class society. These physiological fragments randomly superimposed on Le Brun's drawings, however, mark the furs with the heavy deadness of the fox corpses. The animals' absent life makes itself felt in their three-dimensional corpses. The corpses claim the space between Le Brun's drawings of the fox and his fox-man: if the illusion of three-dimensionality was the purview of the human face in Le Brun's studies of human heads, the three-dimensional animal corpses starkly contrast with the actual flatness of the human face and human gaze. The corpse stands for the only site of resemblance between human and animal. Between two fox muzzles at the top and three fox-man faces at the bottom of the canvas hangs a fox, horizontally. Its lifelessness is the third element—i.e., the third fox—in the comparison.

In addition, the triangular jawbones envision an escape. As famously suggested by Deleuze and Guattari, trapped, domesticated animals (first among them the domesticated, Oedipalized subject) can escape territorialization and take flight.[17] The triangular jawbones also look like arrows pointing in all directions: up, down, inside, outside, and away; they seem to address, that is, animals as precisely those beings that escape immediate visual grasping (i.e., or representation) and that thereby remain outside the frames of human knowledge.[18] They appear as a kind of nonmimetic writing, emblems of some other writing—distinct from sensuous resemblances and the immediate recognition that resemblances produce. As such, they reveal a "hole" in vision: Spoerri's *Carnival* suggests that vision has only been able to produce a series of recognizable images, i.e., fetishes, whether in science, art, or commerce.

Finally, by occupying the blind spot of vision, the jawbones in this assemblage also morph into other incommensurate animals that suddenly and unexpectedly flicker through the still present resemblances. Little eyes are placed on two jawbones in a manner that sketches the profile of two ducks.[19] Again, these ducks undermine the classificatory system of resemblances (ducks equal foxes or ducks differ from foxes) produced by what the philosopher Giorgio Agamben terms "the anthropological machine," a machine that, for Agamben, is always optical because it produces the animal first as the site of visual recognition and then of separation from the human.[20] With their flickering presence within the jawbones, these strange and funny ducks intimate the possibility of unforeseeable, unenvisioned becomings.[21]

Multiplied Wolf-Men: Le Brun, Beuys, Freud, and . . . the Meat Grinder

In another of *Carnival*'s assemblages, Spoerri adds a hanger with a wolf mask and wolf coat to Le Brun's ground depicting the wolf and the wolf-man. Real traps also bite into the fur coat. The coat echoes some mythic hunt while explicitly quoting contemporary art, i.e., art's making and remaking, and its mythmaking characters. Indeed, the wolf-fur coat invokes, somewhat ironically, a legendary figure of contemporary art, Joseph Beuys. The irony lies in Spoerri's remaking of Beuys's legend through Spoerri's own unverifiable account of the wolf coat in the assemblage. Although Spoerri and Beuys were friends, and Beuys collaborated with Spoerri more than once, their art speaks different languages. In *Carnival* the coat exemplifies the "fetish-character" of the name of the father, Beuys: Beuys's name is always and immediately associated with his totemic attire, his famous attachment to felt clothing, to his hat and to his vest. Spoerri oxymoronically invents "ready-made" Beuys, while he also empties the name in *Carnival* through the unworn, hung wolf-fur coat. Beuys's "originality" is lost.

According to Spoerri, he had originally purchased the Siberian wolf-fur coat but then lent it to Beuys. The latter used it in one of his works, and Spoerri was thus never able to retrieve it. Later, Spoerri bought a similar coat, this time made with Alaskan wolf fur, while in New York.[22] The coat thus moves from symbol to commodity to loan to work of art and back again to commodity (or copy). Spoerri used the Alaskan wolf-fur coat in the assemblage under discussion, where the coat is set against the ready-made resemblances of Le Brun, serving as a physical (but still empty) embodiment of Le Brun's wolf-man via its subtle reference to Beuys. The coat suggests at once early culture—the presumed totemic power of wolf fur for early humans—and contemporary "relations" among animals, commerce, and art. Specifically, it juxtaposes Le Brun's resemblances with Beuys's animal performances (impersonations, conversations, etc.) and, more precisely, Beuys's 1974 cohabitation with a coyote, also held in New York (*I Like America and America Likes Me*). In commenting on the coyote action, Beuys explains his work with animals as an attempt at recuperating the power (energy) of the spiritual existence of them all. On his account, he tries to "speak" with animals in order to enter another kingdom, a realm that people have forgotten.[23] He hopes to make contact with America through the reenactment of the traumatic encounter between Native Americans, whom the coyote represents, and the "Western Man." Arguably, Beuys presents himself, the artist, as the shaman endowed with healing powers: his contact with animals, although revealing a traumatic repression, aims at a mystical union, of which he becomes the initiator and medium. Indeed,

32 Daniel Spoerri, *Carnival of Animals: Reproductions of Humans Compared with Those of Wolves,* 1995. Assemblage on enlarged scan, after Charles Le Brun (1670), with one additional object, 140 x 100 cm and 140 x 50 x 50 cm.

Beuys's action exemplifies the unequal relation between human creativity—the human hand and the "impulse of consciousness"—and the un-freedom of the coyote, reduced to total animal energy, which ultimately only the shaman artist can liberate. In this regard, Beuys's communications with animals, and his

33 Daniel Spoerri, *Fleischwolfbrunnen* [Meat-grinder fountain] / *Fontana Tritacarne* (also known as *Gocciolatoio* [Drip stone]), 1961–91. Bronze sculpture, 3.25 m x 2.10 m.

guises, are just another facet of physiognomy, as it filters through Le Brun's drawings. Spoerri appropriates both to counter their effects.

In Spoerri's assemblage, the coat hangs on a rack where, instead of a hat, we find the mask of a wolf, itself in lieu of the wolf's head. The wolf's head, absent from the rack, is represented flat by the drawing on the wall, which is further compared to the head of the wolf-human sketched directly below it. The wolf is chopped into pieces in keeping with both human instrumental reason and commodity-fetishistic desire: indeed, Spoerri bought the coat, while Le Brun, in intellectual fetishistic manner, fragmented his animals and people and focused only on the head, as the most important (expressive) part of the human body in relation to the soul.

Spoerri's wolf fur is then shown as ferociously caught and injured by numerous traps. Propped up vertically as if it were a human figure—its head could be the wolf-human muzzle also depicted by Le Brun and displaced on paper—the wolf is inhumanely wounded, killed, and impaled, as a trophy, first, of commerce and, second, of art. The wolf returns reincarnated grotesquely as human. In order for humans to domesticate their fear and to satiate their own and their civilization's wolfish voracity, it seems that they make themselves into the wolf, which they can admire only as trapped—possessed through the logic of the same—into a "human" or vertical, bipedal stature, itself empty.

Yet the wolf takes its vengeance: Spoerri's wolf-man elicits memories of Sigmund Freud's famous Wolf Man case, only to indict psychoanalytical interpretation. The indictment seems to follow Deleuze and Guattari's critique of psychoanalysis, namely, that psychoanalysis aims to free the human from its "wolf." In Freud's Wolf Man case, for example, the wolf coincided with the father caught in the act of sex by the son, who, in his adult life, suffered from the repression of this event. Freud had hoped to help the son cope with this initial traumatic act of seeing with the aid of psychoanalysis. Yet, as Deleuze and Guattari point out, psychoanalysis only seemingly liberates humans. In fact, it domesticates them, and—in the Wolf Man case—it domesticates the wolf as well.[24] For Spoerri, the human may have temporarily escaped the wolf-man assemblage, yet it has left behind the injured wolf fur: the wolf has definitely been trapped. But Spoerri's trapped and injured wolf (man), whose eyes remain empty in the mask that is on display, may still find its (his) pack by rejecting the human figure/body altogether and fleeing elsewhere.

Indeed, Spoerri exhibited an imposing fountain for the Seville exposition in 2000, entitled *Fleischwolfbrunnen,* which is now on view at his sculpture garden in Seggiano, Italy (fig. 33). For this fountain Spoerri used hundreds of meat-mincing machines. The German word *Fleischwolfbrunnen* is a common one that literally translates as "meat-wolf(ing)" and that denominates this ordinary kitchen tool. Where at first glance the wolf appears to return in the fountain as domesticated meat-mincing machine, at second glance it is liberated from the injuries it suffered in *Carnival.* The substantive *Fleischwolfbrunnen* sounds like a verb in German; in a way, it changes into an action, something like "to fountain mincing-machines." As a meaningless, nonexistent verb, the wolf here multiplies into an incessantly pouring pack of wolves, a pack of wild, devouring animal machines.[25] This pack of wolves gains power and autonomy—freedom—from the human image of the wolf, an image that these wolves do not resemble. The tool displaces its function within the human world, and devours human ideas and the idea of the human itself: all that one can see and recognize in Spoerri's wolf fountain are the human remains of one foot and hand at its base.

In Lieu of a Conclusion: What Has *Carnival* Got to Do with It?

What has the title of Spoerri's series of assemblages, *Carnival of Animals,* to do with the work itself? Where do the "carnivalesque" aspects of animal and human resemblances—so ferociously depicted—emerge in this work? One can read the word "carnival" in a simple, profane way, i.e., as one understands the word in common parlance. One can thus highlight the custom of masking oneself as another, taking up someone else's or something's guise. Masks are often animal masks, whether for purposes of caricature, defilement of authorities, or the incarnation of the totemic power of animals. Not too distant from the popular meanings of carnival is Mikhail Bakhtin's notion. Bakhtin considered carnival a profane union with the cosmos, a ritual of death and resurgence, and a site where tradition meets subversion. Spoerri's *Carnival* is pervaded by mortality. This haunting, however, is not only terrifying; rather, it intimates change and becoming. Death in carnivals hints at this dialectical turn, i.e., the transformation of death into invisible regeneration. It is invisible because regeneration lies beyond the frame of both art and science within which the human imaginaries have been captured. Bakhtin writes that the mask, in the popular festive spectacles of early modernity, "is connected with the joy of change and reincarnation, with gay relativity and with the merry negation of uniformity and similarity."[26] In his view, masks are associated with the grotesque open body and with its regenerating power in folk culture.

Similarly, carnivalesque culture, which Spoerri recycles from popular flea markets, was familiar with the element of terror and fear, Bakhtin points out, "only as represented by comic monsters that were defeated by *laughter.* Terror was turned into something gay and comic." Spoerri's *Carnival,* through its dialectical montages of ground and foreground, shows us the horrors of modern classification and of the mortification of animals and humans. According to Bakhtin, in modernity "the mask acquires a somber hue. A terrible vacuum, a nothingness lurks behind it."[27] Yet *Carnival* does not forget laughter. Note that the laughter it provokes is not that of humans laughing at animals but rather of animals and humans laughing at "serious" human absurdities. About these physiognomic experiments, Spoerri said in an interview with Sandro Parmiggiani, "In my own way, by using found objects, and with much *irony,* despite the fact that I am no scientist, I enjoy developing all these speculations, which were already absurd in earlier times."[28]

But what does Spoerri's *Carnival of Animals* accomplish with its laughter? How does the "carnival" in *Carnival* liberate the animal-human relations from the anthropological machine, here in the guise of physiognomy? First reduced to two dimensions, then trapped in human-animal resemblances, and finally objectified/fetishized in things (or interpretations), Spoerri's carnival animals yet manage to escape science and art—knowledge, the frame. These invisible animals take flight-defying gravity and classification, leaving us with the shells of our own projections. In an "off-screen" space of nonknowledge, Spoerri's *Carnival of Animals* approaches Deleuze and Guattari's idea of art's work. They write that the arts have no other aim than to unleash becomings. Steve Baker thus comments that, in their view, art consists in "letting *fearsome things fly*" (emphasis added).[29] Now, if in *Carnival* the human logic of resemblances appears—for an instant—insignificant so as even to provoke laughter, it is because the artifacts, masks, furs, skulls, and skeletons in it emphasize the gap, the incongruities, that (by chance) both separate us from and unite us with the images of animals that humans themselves have created. These animals are the mirrors and masks of human limitations and fears (of difference and death), which the unenvisioned animals in art throw back at us in their laughter as they take flight from the immobilized, immortalized, and mortified images of themselves they leave behind for us to look at.

While the popular element is certainly important to Spoerri, whose *Carnival* pokes fun at high art (representation, illusionism, etc.) through the willful intrusion of the messy quotidian and kitsch, subversion requires reenacting *ad absurdum* the "visual" commonplaces and disciplinary discourses in which art (objects) and humans (their ideas) are situated. From this viewpoint, if *Carnival* provokes laughter, it is because it breaks "up all the ordered surfaces and all the planes with which we are accustomed to tame the wild profusion of existing things."[30] Michel Foucault laughed for this reason when he first encountered a Chinese classification of animals. Both the laughter and the classification, then, by his own account, constituted the main reasons behind his investigations of human knowledge in *Les mots et les choses.*[31] Through its parodic relation to ingrained structures of seeing, i.e., through the avant-garde *mise-en-abîme* of the logic of resemblance, *Carnival* reenacts and exasperates the carnivalesque dynamics—and black humor—of all avant-garde art, thereby setting free its "invisible" animals.[32]

A Conversation with the Artist Mark Dion

JOAN B. LANDES, PAULA YOUNG LEE, PAUL YOUNGQUIST: *Gorgeous Beasts,* the title of our volume, calls attention to the doubleness of human relations with animals: we are both captivated and repelled by animals. We adopt them as pets, hunt them for food or trophies, collect them in zoos and menageries, and preserve them as taxidermies within natural history collections and private homes. Whether through acts of love or extinction, we constantly impose order on animals in our desire to know them. You have observed elsewhere that the problem is not with taxonomy itself but with the ideologies that infiltrate taxonomy. Can the paradox of *Gorgeous Beasts* offer an entry point into a more critical, but also more playful approach to animal subjects, as you have attempted in your own work?

MARK DION: I know what you mean; our society exhibits such wildly contradictory attitudes toward animals. However, what strikes me is that we continue to regard them as central and important to the human experience. This seems important to me, since so much of our culture devalues anything outside of conventional capital exchange, and wild animals are certainly on the margins of capital.

Insofar as actual contact with living flesh-and-blood animals is no longer a part of the everyday life of many, we have filled that gap with surrogates. Animals continue to be seminal in the stories we tell, and we surround ourselves with beasts in porcelain, bronze, plastic, and plaster. Images of birds, fish, and mammals adorn our wallpaper, flatware, and the pictures that line our walls. What American child's bed is devoid of a teddy bear?

This apparent need for animals in our culture does seem like an opportunity for a more complex and critical dialogue about the place of animals in the world. Certainly, wildlife conservation organizations have been aware of how to exploit the *Gorgeous Beasts* paradox. They have been able to make tigers, sharks, and bears into cute fuzzy icons able to coax dollars from suburban pockets. This concept of the charismatic megafauna has allowed organizations to protect a wide range of organisms and habitats but use a single animal as a poster child. Unfortunately, this does little to broaden public understanding of the complexities of international wildlife conservation. I also employ surrogates in my work to speak about real animals. When using goofy plush animals and even taxidermy antiques, I am employing them like a ventriloquist's dummy to humorously speak grim truth.

JL, PYL, PY: Yes, by speaking grim truths that demand the viewer's attention, as when you present an entomological subject, a mole awkwardly suspended by a rope from a ceiling hook (*Les nécrophores–L'enterrement [Hommage à Jean-Henri Fabre],* 1997), or, with particularly mordant humor, when you encapsulate the tragic history of the bison in a diorama, which is itself mounted on a moving cart (*Mobile Wilderness Unit,* 2001), suggesting that wilderness can now be trucked out when it is needed, even after the disappearance in the wild of an animal or plant species . But equally, in your works, there is always a marvelous element of play, which disturbs but also amuses (for example, *Polar Bears and Toucans [from Amazonas to Svalbard],* 1991). By inviting the laughter that accompanies a child's surprise, you bring us closer to realizing our own animal nature and the animals' place in our species' survival. Your works do indeed raise questions about the limits of conservation as an institutionalized politics, but you don't refrain from asking us to laugh at the absurdity of the way in which the gorgeous, what is beautiful, is tied ineluctably to a long history of political violence between humans (as a natural species) and animals. Can you share your thoughts about the role of laughter in your art?

DION: In a society where hypocrisy, ignorance, cynicism, and self-interest are so much a part of the framework of real politics, I think laughter is one of the only weapons left to us. Is it surprising that most intelligent Americans trust *The Daily Show* and *The Colbert Report* more than the supposed legitimate news networks? At times, parody and a certain Brechtian humor are the only cards we are left to play in the game of resistance. It seems clear to me that with regard to critical environmental issues and wildlife and wild place conservation, we as a society cannot muster the political will to make the fundamental philosophical and practical changes necessary to enact the kind of society which would be able to enact meaningful solutions. Thus, for the things I care about—marine diversity and health, wild places and wild animals—the only possible result is a slow grinding degradation resulting in a less interesting, diverse, and beautiful world. In the face of such a pessimistic future I can only imagine fighting back through humor, which attacks irrationality and hypocrisy, on the one hand, and [promotes] strategies that magnify curiosity and wonder, on the other. Were it not for humor, the only position I could imagine would be a paralyzing melancholy.

With regard to institutional conservation and environmental politics, one just has to laugh at the numerous attempts organizations make to contort preservation and protection of the natural world into a capitalist framework. How many wild schemes they hatch to accommodate the corporate masters of the universe, in comic carbon swaps, for example. Perhaps we should just admit that capitalism is not going to save the world. Perhaps we should admit that capitalism is exactly what the world needs to be saved from. Just because we are "Greens" does not exempt us from boneheaded thinking.

JL, PYL, PY: Yes, we are living through an acceleration of horrors wrought by the excesses of corporate power. Against this, the avatars of each threatened animal compete for public attention, helping us forget that human animals are likewise threatened by the rapacious workings of capital (*Landfill,* 1999–2000). Just today, the Defenders of Wildlife, one of countless environmental groups, called out to save the sea turtles from the consequences of last year's Deepwater Horizon oil disaster. Poignantly, the turtles' defenders felt they had to apologize for making their plea during what other environmentalists had already designated as "Bear Awareness Week." Here is where art as a critical pedagogy might help break through the paralyzing melancholy that accompanies information overload, on the one hand, and, on the other hand, the social psychosis or repetition compulsion writ large, whereby we are asked to accept as the solution to our ecological problems the very same actions that created them in the first place, to wit, your example of carbon swaps. In the face of all this, how can the comic absurdity of finding animals in the museum become a teachable moment?

DION: Art is a pretty tricky teaching tool, just as it is a rather imperfect tool for social activism. Both of those fields imply an honesty and straightforwardness, which you cannot assume of the artist. Artists excel at lying. Also, both traditional teaching and advocacy are based on reliable facts and information, and as an artist I am not necessarily obliged to speak the truth; that is not in the artist's job description. Indeed, the bread and butter of my craft is the implementation of subtle irony, humor, parody, and most of all ambivalence. These are of course also the tricks of a sophisticated, crafty teacher, but they are on the fringe of the norm, whereas they are fundamental to the field of visual art.

I think that art speaks best when it defies the simplicity of right and wrong and attempts to give voice to the complexity of ambivalence, melancholy, mourning, uncertainty, and the complex, irritated state in which we often find ourselves when the contemporary society fails miserably to fulfill its promise of rationality, in other words, the absurd. The artist is comfortable as a voice of uncertainty, and that is a difficult position to articulate in a world craving clarity. I know that Joseph Beuys saw the artist as the mythic figure of the shaman, but I see the artist in oscillation between the trickster and shaman. I don't know about you, but my best teachers were more tricksters than saints or shamans. One thing Beuys seemed correct about is that the artist's totem animal is the rabbit—Bugs Bunny, to be specific.

JL, PYL, PY: What you say calls to mind one of your most memorable pieces: *Survival of the Cutest (Who Gets on the Ark?),* from *Wheelbarrows of Progress* with William Schefferine (1990). A wheelbarrow of stuffed animals defies the sacred rules of Old Testament religion. By substituting toy animals for their counterparts in the biblical story and by playfully asking viewers "who gets to decide?" you cunningly subvert a cultural narrative sanctioning human dominion over animals, still prevalent in much contemporary talk about species protection. Indeed, in many of your thought-provoking works, you have substituted the "artificial" animal—stuffed, plastic, or metallic—for its "real" counterpart. How do these ludic, but also commonplace, animal bodies interpose themselves in the divide humans impose between their human selves and their animal familiars? Does it take a trickster to break us from our habits and gesture toward a relationship to the animal such as a child might have to his/her favorite plushy?

DION: The work you mention was a part of a group Bill Schefferine and I produced called the *Wheelbarrows of Progress*, each of which attempted to discuss some aspect of environmental activism and philosophy fraught with contradiction, from monkey-wrenching strategies to tropical forest conservation. *Survival of the Cutest (Who Gets on the Ark?)* manifested out of the discussions around the notion of "charismatic megafauna" that were openly debated by the major wildlife-protection organizations at the time. Clearly, there are animals that are immensely appealing to a broad public, like pandas and whales and tigers and elephants, and it is easier to raise money for these creatures than, say, snails and pikas and sparrows. Conservation organizations know how to exploit the image of these well-loved beasts, while at the same time they know that saving the habitat for tigers will also function as an umbrella for hundreds of less sexy organisms. Many of the charismatic megafauna are larger animals, which have vast territories, and so saving these poster children also protects the rest. I guess the disappointing aspect of all this is that it would be more straightforward and culturally productive to recognize the importance of saving ecosystems. Some of the animal stars are of course keystone species, meaning that they create habitat for other things. Beavers, corals, and elephants are examples of that, while many of the sexy beasts are apex predators or marginally important for the general ecosystem. To me it seems silly to protect individual animals or species without place. What would the point be if we have tigers only in captivity and no longer in the wild? A zoo animal is a valuable education tool, but it is merely a representation, not unlike the animal in a diorama, photograph, or painting. Wild animals are 90 percent place and 10 percent flesh and blood. If we cannot preserve a place for tigers, we don't deserve them anyway.

So, I guess, to get back to your question, last week I attended a conference on art and taxidermy at the Natural History Museum in London organized by the art historian Petra Lange-Berndt, who is the expert on this issue. I showed slides of my works, which include traditional taxidermy, taxidermy animals in anthropomorphic poses, animals made from other animals (polar bears made of goat skins, for example), realistic animals made of artificial material, and found plush toys. Some of the taxidermy people were upset because I mixed all this together, but to me there was little difference between the various forms of representing animals; there simply was no nature here. This is all trickster territory.

JL, PYL, PY: Taxidermy is a continuing motif and methodology in your works, and, as you say, you have been willing to violate the rules of taxidermy, especially the commandment to make a specimen appear, nay, *be* "natural" (*Ichthyosaur,* 2003). There are so many examples to which we could point, but one that especially intrigues us is the fox in your installation piece *The Delirium of Alfred Russel Wallace* (1994). This fox is lying on its back in a hammock, wearing spectacles, almost smiling as he muses upward—in short, hardly a traditional taxonomic fox pose. Even in traditional taxidermy, let alone the hybridized and inventive versions you integrate in your many works, the line between beauty and the grotesque is very thin. Is this mixing of the visceral and the intellectual a quality that you privilege, or is it a by-product that accompanies any serious inquiry into the human relationship to nature?

DION: I really respond to that notion of mixing the visceral and the intellectual, or rather the idea that the visceral or intuitive can actually be a catalyst or lead one to a thoughtful and critical place. Too often, when thinking about art, people are quick to respond to the formal use of text, or dematerialization of art objects as the only legitimate intellectual art. They are dismissive of martial or visceral strategies, imagining "cool" art as theoretical, while "hot" art is merely expressionistic. That is all a bit too stingy and Protestant for me, as someone who makes meaning with material things and animal bodies. I strongly would like to affirm the notion of a philosophical and intelligent art, which works with a vocabulary of material things and the strategic deployment of intuitive strategies.

One of those strategies is the deliberate use of anthropomorphism, which is employed in the Alfred Russel Wallace piece. To employ anthropomorphism, or allegory, metaphor, or irony, is to drastically break with the vocabulary of science. This is a set of tools utterly unavailable to science, as perhaps the visceral is as well—although I think the role of intuition is underestimated in its application in scientific research. It may be a hindrance and highly problematic to encourage even the slightest whiff of anthropomorphism in the biological sciences, yet I cannot imagine how to practice art, or anthropology for that matter, without knowing its subtle power.

Working with nature in a physical sense can obviously be dangerous territory because of its messy and unpredictable character, both conceptually and actually. There remains so much social discomfort with reminders of our own animality, which of course seem bound to the anxiety around our mortality. Also, there is a practical complexity, in that art making and viewing are largely cosmopolitan endeavors. I recall that when I was formally studying, there was virtually no faculty present at any of the schools I worked at even remotely interested in the politics of the representation of nature. Today I feel remarkably fortunate to find animal studies attaining such heights of scholarly credibility. Perhaps in the recent past ecological and animal issues were viewed by many social critics as entirely bound in wilderness conservation agendas, which in turn were seen as hopelessly romantic and bourgeois. There was very little respect for thinkers who took environmentalism seriously.

JL, PYL, PY: Yes, it's urgent that ecology and animal issues not be confined to wilderness conservation agendas. In so many ways you have boldly disturbed the disciplinary boundaries, not to mention the boundaries of propriety, which serve to cordon off ethical responsibility and action from the practice of science and even the making of art. How often have Natural History Museum visitors, even today, unconsciously accepted the ethical boundaries that silence protest against the hidden assumptions shaping the dioramic display of lost humans and animals? In contrast, in *Black Rhino Head* (1989), and other works from your extinction series, or in your many provocative installation pieces and works on paper, where the animal specimen is covered with black tar, you disturb the comforting containments of ethical response, inviting your viewers to embrace, question, and learn the history of the disappearances that colonialism and capitalist exploitation of the earth have wrought—and perhaps also to examine their own complicity in these actions. The installation piece *Killers Killed* (1994–2007), is especially compelling in this regard. Here you dare to substitute tarred taxidermic animal bodies—and purportedly "guilty ones" at that, e.g., animal killers who have met their just end—for the tortured southern black men who suffered for decades the fate of angry lynch mobs. Like the slave runaways who were subjected to the trained killer dogs of their masters or the innocent victims of lynch mobs, the presumed moral and physical "ugliness" of the guilty prey is undone by the pathos of the hanging tree—making it hard to take comfort in the animal's death.

DION: These works are dark and grim and hard to look at, even for me, the maker. Works in the manner of the grotesque, from Goya, to Cindy Sherman, Joel Peter Witkin, Warhol's "Disaster" works, Jake and Dinos Chapman, John Isaacs and early Damien Hirst, are challenging for viewers. The *Killers*

Killed group of works was largely inspired by an etching from the period of the Thirty Years' War, in which a group of former soldiers-turned-brigands festoon an oak tree, while soldiers and priests mill about the trunk (Jacques Callot, *Der Galgenbaum* [*The Hanging Tree*], 1633). I found this in Munster, where I also was told of how revolutionaries' bodies would be tarred and hung from the cathedral in cages as warnings, just as criminal bodies were tarred and left on gibbets at the crossroads, and pirate corpses were arranged at the harbor's entrance. There is a long history of both hanging and tarring and intolerance, and some perversion of justice needing to be on display. And of course there is the reference to the murderous ignorance of lynching, which victimizes the innocent. The intended display of bodies also links to ranchers' and farmers' practice of hanging shot or trapped pest animals on a line or fence to ward off other members of the same species—coyotes, weasels, crows, etc.

The animals in these works are all those designated as r-selected species, which are usually organisms intentionally or accidentally introduced into a novel environment, which they quickly dominate and drastically alter. They are specialists at nonspecialization and tend to thrive in human-disrupted ecosystems, where they can rush in and dominate a wide variety of niches, pushing indigenous species out. Once established, they are remarkably difficult to eradicate. The zebra mussel, kudzu, starling, brown rat, German cockroach, Asian carp are all typical of this group, as are most of the animals we associate with human habitation. They are detested and persecuted by the society responsible for their introduction. This is of course a global phenomenon, a sort of wildlife roulette, which shapes and forces the hand of evolution. The painter Alexis Rockman and photographer Bob Braine and I did a good deal of collaborative work around this issue in the mid-1990s.

JL, PYL, PY: Yet around this same period of the early 1990s, you produced some of your most evocative works, marked by tenderness, even melancholy. We are thinking of the already mentioned *Black Rhino Head,* as well as *Grotto of the Sleeping Bear—Revisited* (1997). Or another previously mentioned work, the suspended mole in *Les nécrophores-L'enterrement (Hommage à Jean-Henri Fabre).* In each of these displays, the once alive animal reappears as a kind of dream subject. What is more, you often frame the animals—or, in the case of the rhino, animal parts—within the shipping containers that might have delivered them to their final destination. What is missing is the scene of the animal's death: the act of killing is far removed, as are all traces of the animal's fierce response to danger. Can you say something about your mood, as both artist and future spectator, in composing these works, especially in contrast to the grim and hard-edged subjects we have just been discussing?

DION: In some of these works, particularly the sculptures of resting or sleeping bears, the animals are being depicted as thinking/dreaming entities. They are placed into predicaments, atop piles of material culture, in pools of liquid tar and objects, imprisoned in towers or dungeons, and they obviously anthropomorphically reflect on their state. The viewer is forced to ponder the question of what is the animal thinking. Which is a question I ask myself on every single visit to the zoo, or the stable where my wife keeps her horse, or when I encounter a bear or deer at my place in rural Pennsylvania, or when I watch my dog twitch in her sleep. I find this kind of question very rich, and I would like to place my viewer in the situation to have to contemplate it, yet also to direct those thoughts by the tools of the artist, context, composition, materials, form, color, space.

I think that you are accurate when you use the term "melancholy" to describe the tone of these works. They are not invested in the excitement or adrenaline rush of a confrontation with an animal, but rather they are witness to the aftermath, or the viewer is in the role of the voyeur, observing but not possibly participating. Sometimes the animals in these works are meant to be animals, and other times they are meant to be trophies or specimens.

JL, PYL, PY: Speaking of specimens, you recover in your work the marvelous and, from a modern (taxonomic) perspective, marvelously disordered place of the specimen in early modern cabinets of curiosity (*Theatrum Mundi*, 2001). Likewise, the proximity to animals within your reinvented cabinets reattaches viewers to the passions of collectors in the act of collecting, which is after all a kind of property claim, an act of ownership. Viewed from this perspective, many of your installations around the theme of cabinets seem to be a kind of meditation on the double nature of collecting and curiosity, as both loving acts and deadly gestures. Can you say more about the place of the animal in cabinets and libraries that you have invented, as well as your recovery of traditions of specimen keeping and cherishing, which predate the institutional rise of natural history museums?

DION: The animal body in both natural history institutions and in pre-Enlightenment collections has a powerful responsibility. This is perhaps even more true of the animal in Renaissance collections, since it may represent not merely the animal itself, or its place in orthodox taxonomy or some evolutionary scheme, but it may have fetishistic value as a magic object as well. The specimen can possess a very real and practical magical power for the collector, bestowing on him protection or granting health or particular abilities. One of the aspects of these collections, which fascinate me, is that while one can discuss them as the incubators of scientific thought, they can simultaneously be imagined as the end of the hermetic tradition. These collections seem so clearly about an attempt to comprehend the methodology and logic of the godhead through material accumulation and arrangement. That is perhaps not so different from how natural history institutions function, if we change the term "God" for that of "nature."

It is, however, in the complexity of representation itself that I think these collections find sympathy with contemporary art, and sculpture in particular. In the way that an artwork's meaning is not merely an illustration of its maker's concept, these animal specimens are not intended to just signify the animal, and that is one of the complexities I hope to explore when I evoke the Wunderkammer tradition in some of my works. I find tremendous sympathy between "the specimen" and "the sculpture."

JL, PYL, AND PY: And, moving forward, from the Wunderkammer to modern biology, one encounters rules by which the specimen gives its name to the type, the type being an example that serves to anchor the defining features of a particular taxon or set, according to which some organisms are included and others are excluded. The specimen serving as the foundation for taxonomic categorization is what we typically encounter within museum and university collections. As you say, from God to nature: the animal specimens arranged within a collection constitute meaning, which in turn dictates how like and unlike animal objects are arranged. In a marvelous series titled *Extinction, Dinosaurs and Disney: The Desks of Mickey Cuvier* (1990), you portray the great early nineteenth-century French zoologist Baron Georges Cuvier, taxonomist and founder of comparative anatomy, as a Mickey Mouse doll—a kind of recycled "sculpture" from the world of popular culture. Posed before several desks, each representing the naturalist's contributions to natural history, the Mickey doll instantiates the scientist's opposition to evolutionary transformation, which you capture brilliantly in the marvelously punning title of one installation in this series: *The Fixity of a Rodent Species.* In addition, each animated tableau lectures visitors, serving up historical and scientific "truths," as do the animated figures at each of the Disneyland parks—one of which is located not far from the Museum of Natural History in Paris, where Cuvier once held the post of professor of animal anatomy. You created this work in 1990, yet today there are likely more Cuvier/Mickey "anti-evolutionists" than existed some twenty years ago, at least in the United States, where creation museums, Noah's ark theme parks, and anti-evolution school boards and state legislatures oppose the teaching of evolution and contest scientific evidence of the origins of life and global warming. Which leads us back to thinking about what you say about the animal specimen within Renaissance collections: randomly ordered, cherished for their magical powers, and nonetheless incubators of scientific thought. Does the sheer joy and fascination of working with animal bodies help to break through the despair that accompanies the contemplation of the ever-accelerating disappearance of animal life under contemporary conditions? Does the place of the animal specimen in your work, not unlike sculpture, offer viewers the chance to reorder what has been ordered, to inquire anew, and perhaps create new meanings in the process? Maybe even to start to do the work of science outside the laboratory?

DION: There is something extraordinary about working with animal bodies, which are so uncanny in their resistance to decay and their ability to be read as "real." Part of what makes their presence in an artwork so powerful to view is that they are used to produce meaning outside of the conventions in which they are conventionally encountered, which is as scientific specimens, or display elements in scientific education, or as trophies. The animals I employ are making meaning beyond these usual assignments. This is perhaps the space that opens up possibilities for a viewer to imagine an animal outside of the discourses of hunting and science, or trophies and specimens.

Is there a kind of joy to using material like this? Well, yes, for several reasons: first, it always amazing to work with material so finely crafted and cleverly made. Second, these animal objects have a remarkably strong physical sensual quality. However, I would not really say that it is possible to find a buffer for despair through the beauty and physical lushness of the preserved animal body. Rather, prepared animals are a kind of abomination, in that they are like zombies, animated corpses, and yet they are exquisite in their artifice. They attract by their sensuality and repulse in that they are a document of violence toward the living beast. Somehow I am reminded of Walter Benjamin's words: "A historical materialist views [cultural treasures] with cautious detachment. For without exception the cultural treasure he surveys has an origin, which he cannot contemplate without horror. . . . There is no document of civilization which is not at the same time a document of barbarism." This warning would be tattooed across my chest in mirror script so I could fix the thought in my mind each day, were I to enter into that sort of thing. It is my charm against the beguiling power of the rare, curious, and magnificent materials with which I produce my work.

1 Mark Dion, *Les nécrophores–L'enterrement (Hommage à Jean-Henri Fabre),* 1997. Synthetic moleskin, rope, synthetic necrophore beetles, jute rope, eye hooks, books. Installation view from the exhibition Mark Dion: Microcosmographia, South London Gallery, London, 2006.

2 Mark Dion, *Park: Mobile Wilderness Unit,* 2001. Construction trailer, taxidermic European bison, painted backdrop, dirt, leaves, stones, artificial fern and fungi, 290 x 170 x 380 cm.

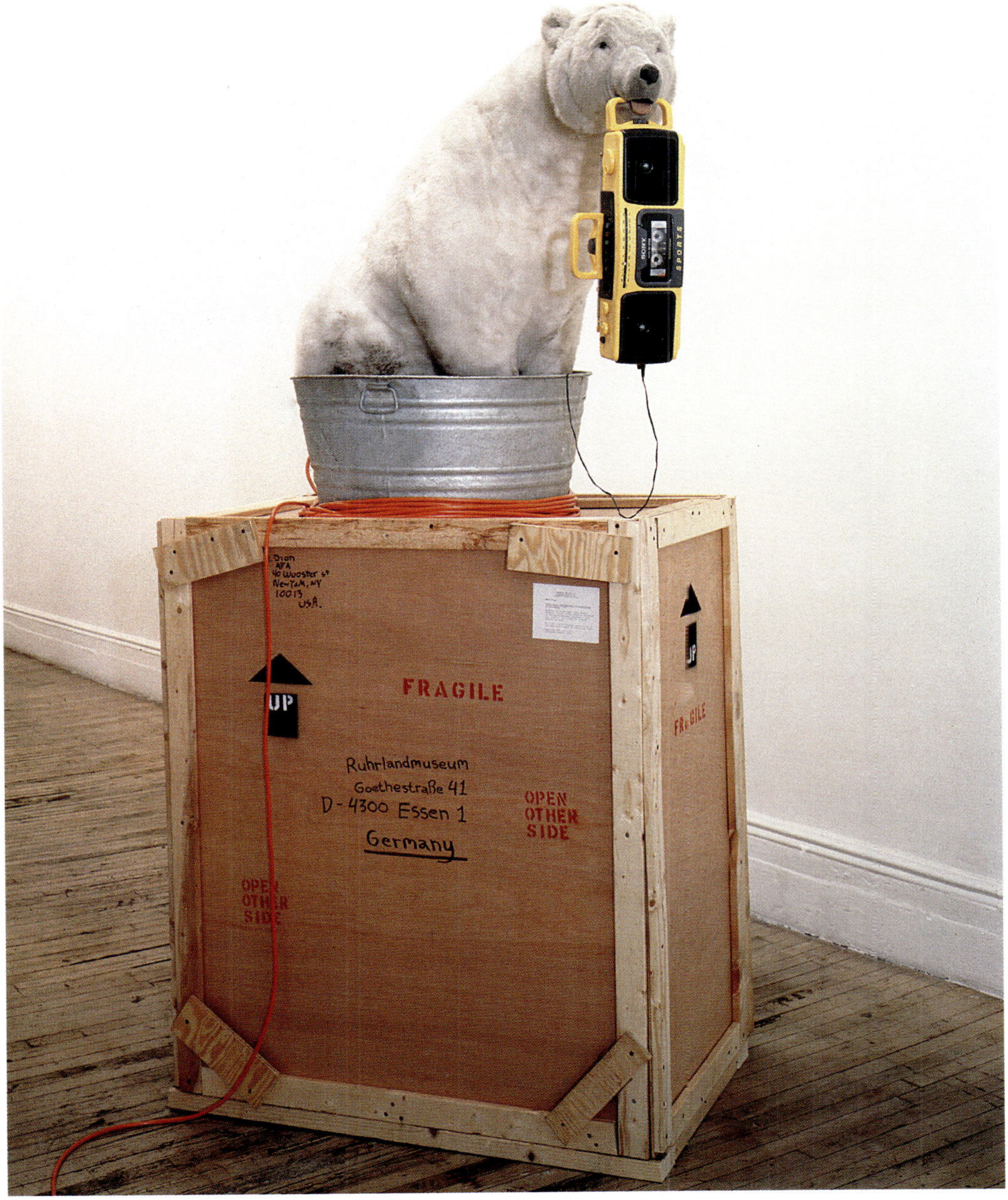

3 Mark Dion, *Polar Bears and Toucans (from Amazonas to Svalbard)*, 1991. Stuffed toy polar bear, Sony sport cassette player, cassette recorded in Venezuela/Amazonas territory, shipping crate, electrical cord, tar, wash tub, 231 x 112 x 75 cm.

4 Mark Dion, *Landfill,* 1999–2000. Mixed media, 71.5 x 147.5 x 64 in.

5 Mark Dion, *Survival of the Cutest (Who Gets on the Ark?),* from the series *Wheelbarrows of Progress,* with William Schefferine, 1990. Toy stuffed animals, white enamel on red steel, wood and rubber wheelbarrow, 63.5 x 68.5 x 141 cm.

6 Mark Dion, *Ichthyosaur,* 2003. Mixed media.

7 Mark Dion, *The Delirium of Alfred Russel Wallace,* 1994 (detail). Installation, mixed media.

8 Mark Dion, *Extinction Series: Black Rhino Head,* 1989 (detail). Wooden crates, stenciled lettering, color photographs, rhino head, wood chips, map of Africa.

9 Mark Dion, *Killers Killed,* 1994–2007. Tree, taxidermic animals, tar, galvanized aluminum, foam, paint, 128 x 65 x 60 in.

10 Mark Dion, *Grotto of the Sleeping Bear—Revisited*, 1997. Synthetic bear, plastic, bubble wrap, wood crate, sawhorses.

11 Mark Dion, *Theatrum Mundi,* 2001. Mixed media.

12 Mark Dion, *The Fixity of a Rodent Species,* 1990. Desk, animated Mickey Mouse figurine, blackboard, curtain, animal skull, nineteenth-century etchings, bird skeleton, books, toys, pointer stick, specimens, eggs, teacup, *Tyrannosaurus rex* models, dissecting kit, audio tape, 150 x 300 x 150 cm.

notes

INTRODUCTION

1. For a deeper treatment of the representation of animals, see Rothfels, *Representing Animals,* and Baker, *Picturing the Beast.*
2. These difficulties are not confined to animals but constitute fundamental problems of semiotics. See Saussure, *Writings in General Linguistics.* On the semiotics of popular culture, see Barthes, *Mythologies* and *Elements of Semiology.*
3. Dion et al., *Mark Dion,* 127.
4. On the relationship between desire, perception, and culture, see Žižek, *Looking Awry* and *Interrogating the Real.*
5. On the various ways in which animals get drawn by and into history, see Kaloff, *Looking at Animals;* Lee, *Meat, Modernity;* and La Capra, *History and Its Limits.*
6. For some essential examples of the developing historical scholarship on animals and society, see Ritvo, *Animal Estate;* Kete, *Beast in the Boudoir;* Robbins, *Elephant Slaves and Pampered Parrots;* Fudge, *Perceiving Animals;* and Henninger-Voss, *Animals in Human Histories.* On the wider issue of the animal in changing attitudes toward nature in the West, see Thomas, *Man and the Natural World.*
7. McShane and Tarr, *Horse in the City,* 1.
8. For more on companion species, see Haraway, *Companion Species Manifesto* and *When Species Meet.*
9. E-mail correspondence of 2 June 2010 with the artist, who also notes that some of the scenes drew from local newspaper accounts.
10. Gombrich, *Art and Illusion,* 198.
11. "It goes without saying that seeing a lion and being frightened of it at the same time, is different from merely seeing it." Descartes, "René Descartes' Sixth Reply." On the beetle in a box, see Wittgenstein, *Philosophical Investigations,* 293.
12. Deleuze and Guattari, *Thousand Plateaus,* 232; Deleuze, "Body, Meat, and Spirit." In the visual arts, see Thompson, *Becoming Animal;* for theoretical approaches to animal studies, see Wolfe, *Zoontologies.*
13. On this theme, see one of the artist's most recent collaborations, Dion et al., *Marvelous Museum.*
14. "Killer Whale Kills Trainer at SeaWorld as Horrified Spectators Watch," 24 February 2010, http://www.naplesnews.com/news/2010/feb/24/killer-whale-kills-trainer-seaworld-dawn-brancheau/ (accessed 19 May 2010).
15. "SeaWorld Orlando to Keep Killer Whale That Killed Trainer," *Los Angeles Times,* 26 February 2010, http://articles.latimes.com/2010/feb/26/nation/la-na-seaworld-death26-2010feb26 (accessed 5 May 2010); "Tilikum, Orca That Killed SeaWorld Orlando Trainer, Will Remain at the Park Despite Calls to Free Him," *Los Angeles Times,* 25 February 2010, http://latimesblogs

.latimes.com/unleashed/2010/02/tilikum-orca-killer-whale-sea-world-orlando.html (accessed 19 May 2010).

16. "Whale Kills Trainer at Florida's SeaWorld," 24 February 2010, http://www.reuters.com/article/2010/02/24/usa-whale-idUSN2421601020100224 (accessed 19 May 2010).
17. See Kathryn Taubert, "Sea World Tragedy: A Former Marine Mammal Trainer Speaks," http://blogs.naplesnews.com/lifeslowlane/2010/02/sea-world-tragedy-a-former-marine-mammal-trainer-speaks.html (accessed 19 May 2010).
18. For the comments summarized here, see the articles cited in notes 14–16.

CHAPTER 1

1. Buffon's *Histoire naturelle, générale et particulière, avec la description du Cabinet du roi* was published in quarto in thirty-six volumes between 1749 and 1789. Other printings and editions soon followed, ranging in size from folio to duodecimo, alongside translations into German, Dutch, English, Italian, and Spanish in Buffon's lifetime (Loveland, *Rhetoric and Natural History,* 10). The *Histoire naturelle* was a highly collaborative project, although Buffon directed and edited it. The first and most famous of Buffon's collaborators was Louis-Jean-Marie Daubenton, who worked on the first fifteen volumes. Along with other textual collaborators, Buffon relied on observations and anecdotes offered by hundreds of correspondents, as well as the contributions of artists, most significantly the illustrator Jacques de Sève. When citing the French edition, I have used the standard forty-four-volume edition of printed text and images. The names of de Sève and other artists and engravers, including [Juste] Chevillet, Jean Charles Bacquoy, Jean-Baptiste Oudry, and A.-J. Defehrt, are given alongside Buffon's in the credits for the illustrations in this chapter. On visual imagery in Buffon's work, see Liebman, "Painting Natures."
2. See Picasso and Buffon, *40 dessins en marge du Buffon.* See also Picasso, *Histoire naturelle.* On Picasso and Maar, see Baldassari's insightful *Picasso: Life with Dora Maar.*
3. Sirens, like harpies, are variously composed of women and birds—sometimes, as in Greek art, as birds with large women's heads, bird feathers, and scaly feet, or as females with the legs of birds, with or without wings. Like the songs of birds, their voices attract, while their bodies seduce. Sirens are also represented as mermaids, as reflected in the fact that the word for "mermaid" in Spanish, French, Italian, Polish, Romanian, and Portuguese is *sirena, sirène, sirena, syrena, sirenă,* and *sereia,* respectively.
4. The other faces seem to draw upon the tradition of *têtes d'expression* and theatrical masks.
5. Symmes, "Illustrated Books," 52, 59.
6. Cox and Povey, *Picasso Bestiary,* 21, 23. Garvey calls "the Buffon . . . one of the greatest Bestiaries of the 20th

century." Garvey, *Artist and the Book,* 158.

7. Cox and Povey, *Picasso Bestiary,* 23. Horodisch suggests that the seemingly peculiar practice of producing the plates and then searching for a text to match has a history as old as the printed books (*Picasso as a Book Artist,* 55, cited in ibid., 195). See also Goeppert et al., *Pablo Picasso,* 104–7, and Matarasso and Benoit, *Bibliographie des livres illustrés.*
8. See, for example, *La Langouste* in Picasso, *Histoire naturelle.*
9. Steven A. Nash, "Introduction: Picasso, War, and Art," in Picasso et al., *Picasso and the War Years,* 21. See *Sheep's Skull* (1 October 1939, oil and India ink on paper, Musée Picasso, Paris); *Flayed Head of a Sheep* (4 October 1939, oil on canvas, Musée des beaux-arts, Lyon); *Still Life with Sheep's Skull* (6 October 1939, oil on canvas, private collection), cat. nos. 34–36.
10. Buffon's outline was announced in the *Journal des Sçavans* (October 1748): 639–40.
11. Dance, *Art of Natural History,* 16.
12. Loveland, *Rhetoric and Natural History,* 10n37. See *Mercure de France* (October 1748): 161.
13. See Gessner, *Icones animalium.* This is an abbreviated version of Gessner's great four-volume *Historiae animalium* (1551–58), a lavishly illustrated encyclopedia of all the known animals, including mammals, amphibians, birds, and reptiles, in alphabetical order, and in various vernaculars, that ran to more than four thousand pages. (The fifth, incomplete volume was published in 1587, after Gessner's death.) The second edition of *Icones animalium* is a large-format 134-page volume, containing a selection of hand-colored woodcuts and descriptions of animals from *Historiae animalium,* but in systematic order.
14. Harrison, "Reading Vital Signs," 189. Yet Harrison rightly points out that, "to a degree, more active experimental practices arose out of attempts to emulate the actual methods of the ancients rather than merely rehearsing their findings." See also Brian Cummings's account of Gessner's recording of claims regarding elephant speech and writing, and his "interest in the instrumental contingencies of the elephant's abilities. The elephant must be taught by a teacher who is physically present, to train his tongue or trunk in the right direction. Then, after he has mastered the rudiments, he is ready to write from a crib, like any apprentice or child." Cummings, "Pliny's Literate Elephant," 166–67.
15. On monstrosity, see Daston and Park, *Wonders and the Order of Nature;* Knoppers and Landes, *Monstrous Bodies/Political Monstrosities.*
16. On this and other examples, see Dance, *Art of Natural History,* 50 and passim.
17. Buffon added, "The plan of his work is good, his distributions show discretion, the divisions are well demarcated, and his descriptions are quite exact—monotonous, to

be sure, but accurate." "'Initial Discourse,'" 158.

18. Interestingly, Claude Perrault was the brother of Charles Perrault, the fabulist (fairy tale author) and designer of the labyrinth at Louis XIV's Versailles.
19. Senior, "Ménagerie and Labyrinthe," 218.
20. In line with the "social" emphasis in contemporary studies of the history of science, Anita Guerrini contests Perrault's claims of empiricism, naïve observation, and absolute verisimilitude, insisting that the animals therein were artifacts, "crafted objects rather than natural ones." Guerrini, "'Virtual Menagerie,'" 30. On animal experimentation, see Guerrini, *Experimenting with Humans and Animals.*
21. Buffon, "'Initial Discourse,'" 158–59.
22. Roger, *Buffon,* 152.
23. Buffon, "'Initial Discourse,'" 167, 170.
24. On Buffon and Locke, see Roger, *Buffon,* esp. 82–83, and Reill, *Vitalizing Nature in the Enlightenment,* 53–54. For an insightful account of "sensationism," see O'Neal, *Authority of Experience.* O'Neal points out the influence not only of Locke but also of Aristotle's concept of the sensory origin of ideas: "his important dictum, rendered usually in Latin by his successors, was: '*nihil is in intellectu quod non prius fuerit in sensu*' (nothing is in the mind that has not first been in the senses, or translated less literally, all knowledge comes to us through the senses)" (2n2).
25. Roger, *Buffon,* 83.
26. Buffon, "'Initial Discourse,'" 171–72.
27. Ibid., 168. He also stated, "We reproach the Ancients for not having constructed systems, and the Moderns believe themselves to be quite a bit above them because they have constructed a great number of methodological arrangements and these lists of classified objects of which we have just spoken. They have convinced themselves that that alone is sufficient to prove that the Ancients did not have nearly the knowledge of natural history that we have. However, the complete opposite of this is the case. . . . The Ancients were far more advanced and knowledgeable than we are, not in physics, but in the natural history of animals and minerals" (166).
28. "Aristotle begins his history of animals by establishing the general differences and resemblances between various kinds of animals. Instead of dividing them on the basis of small special characteristics such as the Moderns do, he gathers historically all the facts and all the observations which bear on the general resemblances and the sensible characteristics. . . . He draws these characteristics from the form, color, size, and all the exterior qualities of the whole animal, as well as from the number and position of its organs, from the size, movement, and form of its limbs, and from the likenesses or dissimilarities which are found in comparison of these

same parts. And he everywhere gives examples in order to make himself better understood. He also considers differences among animals in their style of life, their actions and their habits, their places of habitation, etc. . . . He begins with man, and describes him first, rather because he is the best-known animal than because he is the most perfect." Ibid., 168–69.

29. Ibid., 160.
30. Ibid., 152.
31. Reill, *Vitalizing Nature in the Enlightenment,* 48. For a persuasive reading of Buffon's place within a cross-national vitalist movement in the eighteenth century, see chapter 1, "Storming 'the Temple of Error,'" 33–70.
32. Siding with Philip Sloan rather than with Roger in this instance, Reill explains, "Both Roger and Sloan see Buffon as a leader in formulating a new scientific ideal. However, they differ strongly about the sources of Buffon's inspiration and the degree to which he altered his views over time. Roger saw Buffon as the central figure in proposing a new vitalistic vision of the life sciences that was elaborated by Diderot and the Montpellier school. However, he considered Buffon to have been a Lockean and argued that over time Buffon changed his views considerably, especially concerning the issues of the existence of classes, orders, and *genera.* Sloan is more radical in his conclusions. He denies the direct connection between Locke and Buffon and argues that Buffon was consistent in his formulations. Although I have been greatly influenced by Roger's writings, I believe Sloan's argument is more convincing." Ibid., 273n63. See also Sloan, "Buffon-Linnaeus Controversy"; Sloan, "From Logical Universals to Historical Individuals"; Sloan, "Gaze of Natural History"; Sloan and Lyon, "Introduction," in Sloan and Lyon, *From Natural History to the History of Nature,* 1–32; Loveland, *Rhetoric and Natural History;* Roger, *Sciences de la vie dans la pensée française,* 527–84; and Roger, *Buffon,* 84.
33. Regarding Buffon's repudiation of mechanism, see Reill, *Vitalizing Nature in the Enlightenment,* 33–70.
34. Roger, *Buffon,* 87.
35. Buffon, "'Initial Discourse,'" 161–62.
36. Up-to-date zoological taxonomy is, of course, quite different from that presented in Buffon's work. Buffon groups the domesticated horse, for example, together with zebras, donkeys, and the African wild ass.
37. Buffon, *Natural History, General and Particular,* 3:500.
38. Ibid., 4:242.
39. Hervieux's title is cited in Robbins, *Elephant Slaves and Pampered Parrots,* 124. René Démoris has given the most insightful account of the significance of these mechanical bird organs in the work of the painter Jean-Baptiste-Siméon Chardin. See Démoris, "Chardin, la machine et l'oiseau" and "Inside/Interiors."
40. Robbins, *Elephant Slaves and Pampered Parrots,* 25.
41. Buffon quoted in ibid., 27–28.

42. Ibid., 25–26, 130.
43. William Wordsworth, "The Tables Turned" (1798), in *Poems of Sentiment and Reflection,* quoted in Donald, *Picturing Animals in Britain,* 31–32. Donald objects to Wordsworth's indictment: "Yet the tenor of eighteenth-century natural history writing seldom expressed pride in superior human knowledge or in control of other species. Rather, naturalists were humbled by their own bafflement in trying to order and make sense of the extraordinarily varied and intricate forms of animal life and behaviour. Locke believed 'There is not so contemptible a Plant or Animal, that does not confound the most inlarged Understanding'" (32).
44. Linnaeus quoted in Freedberg, *Eye of the Lynx,* 412–13.
45. On this point I am more persuaded by Alex Potts and Elizabeth Liebman than by Louise Lippincott and Andreas Blühm. The latter stress Buffon's ambivalence toward the image, even his reluctance to use images. "The irony," they write, "is that the illustrations from his multi-volume *Histoire naturelle* are still admired and reproduced, while the text is hopelessly outdated and of interest to the specialized historian alone." Lippincott and Blühm, *Fierce Friends,* 16. See also Potts, "Natural Order," and, for an extended account of Buffon's illustrators, Liebman, "Painting Natures." Liebman makes a convincing argument for Buffon's involvement in the project of illustration of his great tome. On the basis of "Buffon's direct and sustained contact with the artists," she argues, "artistic labor completed his material practice in natural history" (viii).
46. Buffon, *Histoire naturelle,* 1:21–22 and 4:116, 120, quoted in Potts, "Natural Order," 22.
47. Potts, "Natural Order," 22.
48. On the comparison between neoclassical art and scientific illustration, see especially ibid.; Lippincott and Blühm, *Fierce Friends,* 16–17.
49. Potts, "Natural Order," 23.
50. For example, in their study of artists and animals, Lippincott and Blühm remark, "Because the main purpose of images was to record salient details for identification purposes and to permit comparisons over a distance, today they seem rigidly schematic." *Fierce Friends,* 17.
51. For two examples, see the representations of a bear skeleton and *Dere iltis* (polecat).
52. Potts, "Natural Order," 21.
53. Ibid., 22. See also Liebman, "Painting Natures."
54. Cox and Povey, *Picasso Bestiary,* 169, quoting Picasso, *Histoire naturelle,* "le singe," 88 (translated by Cox and Povey). This image is sometimes labeled a baboon, a taxonomically African and Arabian Old World monkey belonging to the genus *Papio,* part of the subfamily Cercopithecinae.
55. Mourlot, *Picasso Lithographs,* 4.
56. I am grateful to Jeff Loveland for his insight on these two issues.
57. Lavin, "Picasso's Bull(s)," 84.

58. This phrase is Lavin's. Ibid., 87.
59. Reill, *Vitalizing Nature in the Enlightenment,* 53.
60. Quoted in ibid., 53. "This assertion," Reill writes, "undercut the mechanist epistemological basis for a universal natural philosophy by closing the breach between mind and matter, man and animal. It also seemed to imply an even more drastic restriction of natural philosophy's truth claims, for it made the contingent the prime area of inquiry."
61. Ibid. Reill observes that "for Buffon this apparent loss was a gain." Striving toward a type of understanding more certain and deeper than that tied to superficial surface phenomena, Buffon's vision of a new science is not unlike Vico's, "even though both began with different premises." For Reill, "they transformed what the mechanical philosophers of nature had considered a mistaken premise—that humans were immersed in a natural world from which they cannot escape—and made it into a virtue, believing that humans were better able to understand this world because they are a part of it, in fact, the most complete part because humans can reflect upon their experiences through the medium of language. But Buffon's formulation was more radical than Vico's '*Verum + factum.*' Vico limited his type of knowledge to human history, whereas Buffon extended it to all living nature, which, in his view had a history similar to that of humanity" (53–54).
62. Lavin, "Picasso's Bull(s)," 79, 80–81.

CHAPTER 2

1. "Absinthe" refers to the drink only after 1842.
2. This making the animal invisible in vellum was challenged, briefly, in the exhibition Fra Angelico to Leonardo: Italian Renaissance Drawings (British Museum, 22 April–25 July 2010). A piece of sheep vellum was displayed next to a piece of paper, and viewers were invited to feel the difference. The curator thus drew our attention to the animal skin that underpinned some of the drawings.
3. Jones and Stallybrass, *Renaissance Clothing;* Stallybrass, "Worn Worlds."
4. For some discussions of the shift from subject to object in Renaissance studies, see Bruster, *Shakespeare and the Question of Culture,* and Harris, *Untimely Matter in the Time of Shakespeare.*
5. A good example of such work is Anderson, *Creatures of Empire.* In this study she describes cattle as "agents of empire" (210).
6. As Steven Connor, reading Michel Serres, has put the relationship between subject and object, they "enter into each other's composition, such that [their] reciprocal constitution . . . is both inaugural and ongoing." Connor, "Thinking Things," 4.
7. The absenting of animals from the animal product, meat, and its implications for the eating of animals is a focus of Carol J. Adams's influential study *Sexual Politics of Meat.*

8. ANT's possibilities for animal studies are outlined in Philo and Wilbert, "Animal Spaces, Beastly Places."
9. In ANT the term "nonhuman" includes animals, technology, architecture, clothing, etc.—all that is not human, in fact.
10. Law, "Notes on the Theory of the Actor-Network," 384.
11. Latour, "Morality and Technology," 247.
12. Bowden, *Wool Trade in Tudor and Stuart England,* 8.
13. Introna, "Ethics and the Speaking of Things," 399. Also working from within ANT, Vinciane Despret has argued that even ethologists—those studying animal behavior—have ignored the sheepness of sheep. See Despret, "Sheep Do Have Opinions."
14. Julian Yates outlines a difference between ANT and thing theory in his article "Accidental Shakespeare," where he argues that "one of the advantages [that the] network-based model of description has over the recent turn back to 'things' in Renaissance studies, and in cultural studies more generally, is that it has no truck with distinctions between nature/culture, animal/human, human/machine that frequently, despite our best efforts, tend to remain fairly stable" (92). Part of what I am arguing in this chapter is that the implied criticism of the turn to things need not be true.
15. Brown, "Thing Theory," 4. See also Brown, "How to Do Things with Things." The Heideggerian underpinnings of thing theory are outlined in Introna, "Ethics and the Speaking of Things," 403–5.
16. Clutton-Brock, *Domesticated Animals from Early Times,* 149.
17. Stallybrass and Jones, "Fetishizing the Glove," 129. Pérez's letter quoted on 128–29.
18. Stallybrass and Jones's focus on the gloves rather than the dog skin could be read as underlining Yates's point, in "Accidental Shakespeare," that material studies assume the human-animal boundary to be a stable one.
19. Berry, *Shakespeare and the Hunt,* 17.
20. On the different kinds of animal skin in use in the early modern period and the methods of preparation, see Clarkson, "Organization of the English Leather Industry." On the representation of the role of the flayed animal skin in disrupting human status in medieval writing, see Kay, "Legible Skins"; I owe this reference to Cary Wolfe.
21. Stubbes, *Anatomie of Abuses,* sig. Qvi^{r}.
22. Shakespeare, *Merchant of Venice,* 1.2.122–27.
23. Stallybrass and Jones, "Fetishizing the Glove," 123. The comment is made in a discussion of Titian's painting *Man with a Ripped Glove* (c. 1523).
24. Jones and Stallybrass, *Renaissance Clothing,* 270.
25. Stubbes, *Anatomie of Abuses,* sig. Cv^{r}.
26. Anston Bosman points out that Stubbes seemed to change his mind when, in *The Second Part of the Anatomie of Abuses, Conteining the Display of Corruptions* (also 1583), he "drops the notion of

leather's primitive simplicity and instead associates it with forms of worldly perversion," because of the corruption associated with leather production. Bosman, "Shakespeare in Leather," 235. I owe this reference to Jonathan Hope.

27. Anonymous, *Newe mery and wittie Comedie,* 4.8, unpaginated.
28. Calvin, *Commentarie of Iohn Caluine,* 569.
29. Anonymous, *Newe mery and wittie Comedie,* prologue, unpaginated.
30. Prynne, *Histrio-Mastix,* 891.
31. On Prynne's ears being clipped as punishment for publishing *Histrio-Mastix,* see Gardiner, *Documents Relating to the Proceedings Against William Prynne,* 17.
32. George Wither, *A Collection of Emblemes* (1635), book 2, illustration 50, cited in Shannon, "Poor, Bare, Forked," 186. Shannon's essay provides a detailed account of the implications of the need to dress humans in animal skins and its negative impact on human status. As I have done, Shannon uses *King Lear* as a text that illustrates and reflects these ideas.
33. See Fudge, "Saying Nothing Concerning the Same," in Fudge, *Renaissance Beasts.*
34. Johnson, *Cornucopiae,* sig. D3v.
35. Dannenfeldt, "Europe Discovers Civet Cats," 404–5.
36. London Metropolitan Archive, MS 9172/33, 119 (24 July 1622). The value of the civet cat lies in its continuing productivity, not its slaughter. Civet could be "harvested" from the animal as often as every two days during the summer months. See Dannenfeldt, "Europe Discovers Civet Cats," 420.
37. Corbin, *Foul and the Fragrant,* 73, 76.
38. Stubbes, *Anatomie of Abuses,* sig. G.
39. Margolies, "Vagueness Gridlocked," 112.
40. Max Horkheimer and Theodor Adorno, *Dialectic of Enlightenment,* quoted in Lippit, *Electric Animal,* 123. Lippit offers a brief overview of philosophical conceptions of smell and the human-animal boundary (121–27).
41. Stewart, "Remembering the Senses," 62. In her essay in the same volume, Carla Mazzio notes that in the early modern period touch fell outside the ways in which the other senses were understood and was thus marginalized: "If touch resists quantification, how could it possibly measure up to the standards of 'rational' inquiry that emerged during the Renaissance?" Mazzio, "Senses Divided," 92.
42. Babington, *Certaine plaine, Briefe, and Comfortable Notes,* fol. 111v.
43. See Shannon, "Poor, Bare, Forked," passim.
44. Jones and Stallybrass note this biographical fact in *Renaissance Clothing,* 180. Bosman traces the meanings of leatherworking in the period and its general implications for a reading of Shakespeare in "Shakespeare in Leather."
45. See Neill, "'Amphitheaters in the Body,'" 28–29.
46. Jones and Stallybrass, *Renaissance Clothing,* 32. The use of the term

"Lendings" is also discussed by Shannon in "Poor, Bare, Forked," 196, and by Bell in "Naked Lear," 55–57. Bell argues that "Lear's declaration about 'lendings' is a philosophical statement (with a postmodern sceptical ring) of the way personal identity is something borrowed, something acquired from the outside" (58). She doesn't comment on the fact that this "outside" might be animal.

47. For an account of the various meanings of "unaccommodated" in *King Lear,* see Keefer, "Accommodation and Synecdoche."
48. On this, see also Shannon, "Poor, Bare, Forked," 196.
49. Hadfield, "Plants in *King Lear,*" 385–86.
50. Topsell, *Historie of Fovre-Footed Beastes,* 758.
51. Ibn Baithar (the Arab record) and Castelli both quoted in Dannenfeldt, "Europe Discovers Civet Cats," 406, 427.
52. Donne, "To Sir Edward Herbert, at Juliers," 218.
53. Louise Hill Curth notes that while parts of animals were used to treat humans, she was "unable to find any examples . . . of parts of humans being used to treat animals." Curth, *Care of the Brute Beast,* 129.
54. Thomas Dawson, *The Booke of Carving and Sewing,* cited in Fudge, "Saying Nothing Concerning the Same," 76.
55. Igor Kopytoff has proposed that even the traded object—the commodity—should be considered to have a biography; for him, the slave is the marker of this. Kopytoff, "Cultural Biography of Things," 65.

CHAPTER 3

1. The epigraphs to this chapter are taken from Césaire, *Discourse on Colonialism,* 34, and Haraway, *Companion Species Manifesto,* 13. All translations in this chapter are my own unless otherwise noted.
2. Agamben, *Open,* 26.
3. Ibid., 26, 37.
4. [Edwards], *Proceedings of the Governor and Assembly of Jamaica,* i–ii.
5. Ibid., iii.
6. The sixth article of the 1739 treaty states "that the said captain Cudjoe, and his successors, do use their best endeavours to take, kill, suppress, or destroy, either by themselves, or jointly with any other number of men commanded on that service by his excellency the governor or commander in chief for the time being, all rebels wheresoever they be throughout this island, unless they submit to the same terms of accommodation granted to captain Cudjoe, and his successors." The seventh article requires that Cudjoe and his forces defend Jamaica in the event of invasion. The terms of the treaty enlist the Maroons in colonial service as a police force and auxiliary militia. See *Laws of Jamaica, 1681–1758,* 258.
7. See in particular Peter Laslett's introduction to his edition of Locke's *Two Treatises of Government;* Michael, "Locke's *Second Treatise*"; Ivison, "Locke, Liberalism, and

Empire"; Lebovics, *Imperialism and the Corruption of Democracies,* chapter 5; Armitage, "John Locke, Carolina, and the *Two Treatises of Government*"; Bernasconi and Mann, "Contradictions of Racism"; and Farr, "Locke, Natural Law, and New World Slavery." For a good introduction to Locke's political theory, see Simmons, *On the Edge of Anarchy.*

8. Locke, *Political Essays,* 179–80.
9. See especially "Of Conquest," chapter 16 of Locke's second treatise.
10. Farr, "Locke, Natural Law, and New World Slavery," 180.
11. Locke, *Two Treatises,* 276–77.
12. Ibid., 272, 273.
13. Ibid., 297.
14. The treaty stipulated that criminal and civil offenses of whites against Maroons be tried in Jamaican courts. Cudjoe and his successors retained the right to punish their own people. See *Laws of Jamaica, 1681–1758,* 259.
15. Balcarres threw everything he had at the Maroons, and then some. He mustered Jamaican militia, dispatched the 18th and 20th Regiments of Light Dragoons, and requisitioned the 83rd Regiment of Foot, then at sea en route to St. Domingo, forcing its return to help stamp out the rebellion.
16. Dallas, *History of the Maroons,* 1:123. Hereafter cited parenthetically in the text.
17. Dallas adds, "The chief motives by which the Council of War were influenced, must undoubtedly have arisen from the apprehension of a general insurrection among slaves on revolutionary principles." Ibid., 1:170.
18. [Edwards], introduction to *Proceedings of the Governor and Assembly of Jamaica,* lxxix.
19. For a discussion of the British deployment of bloodhounds against the Maroons in the context of colonial domination and hemispheric hegemony, see Johnson, "'Give Them Blacks to Eat.'"
20. Rainsford, *Historical Account of the Black Empire of Hayti,* 6.
21. Ibid., 426–27.
22. Ibid., 427.
23. [Edwards], introduction to *Proceedings of the Governor and Assembly of Jamaica,* iv, xxx, xxxiv, lxi, lxx, lxxix.
24. Ibid., xlvi.
25. "Parliamentary Intelligence: House of Commons; Blood Hounds in Jamaica," *Times* (London), 22 March 1796.
26. [Edwards], introduction to *Proceedings of the Governor and Assembly of Jamaica,* lxxix.
27. Rainsford, *Historical Account of the Black Empire of Hayti,* 339.
28. [Edwards], *Proceedings of the Governor and Assembly of Jamaica,* 109.
29. Rainsford, *Historical Account of the Black Empire of Hayti,* 337n.
30. Quoted in the *Times* article "Parliamentary Intelligence."
31. Rainsford, *Historical Account of the Black Empire of Hayti,* 429. Rainsford is willfully disparaging the French here, but for engaging in a military practice—using bloodhounds—for which he earlier faults his British comrades. State-sponsored terror on

the battlefield offends his integrity as a soldier.

32. U.S. Army, *Army Regulation 190-12* (1993), 1. Hereafter cited parenthetically in the text.
33. U.S. Army, *Army Regulation 190-14* (1993), 3.
34. Danner, "Abu Ghraib: The Hidden Story."
35. Sanchez, "Memorandum for Combined Joint Task Force Seven."
36. Agamben, *Open,* 80.

CHAPTER 4

1. Santaniello, "Introduction," in Pyne, *Microcosm,* 1:8.
2. This is in distinct contrast to the work of the twenty-first-century artist Adel Abdessemed, who in *Don't Trust Me* (2008) shows a number of livestock being killed by a sledgehammer to the head.
3. C. Gray, textual commentary, in Pyne, *Microcosm,* 2:184.
4. Wolfe, "Exposures," 8.
5. Quoted in Neeson, *Commoners,* 20.
6. Gray, textual commentary, in Pyne, *Microcosm,* 2:185.
7. There is one other graphic image of animal death in Pyne's *Microcosm:* the slaughter of a hog (2:166, 241). Pyne's text omits any discussion of the image and provides only the simplest of labels: "Cottage Groups: 1. Travelling knife-grinder and tinker. 2. Ditto. 3. Killing a hog. 4. Scalding a hog, and scraping off the hair."
8. Bloomfield, *Farmer's Boy,* 85, 14. Hereafter cited parenthetically in the text.
9. Landry, "Georgic Ecology," 254.
10. Cousin, *Short Biographical Dictionary,* 83.
11. McEarthron, *Nineteenth-Century English Labouring Class Poets,* 1:71–72.
12. Bachelor, *Village Scenes,* 75. Hereafter cited parenthetically in the text.
13. Youatt, *Cattle,* 4.
14. Carter, *His Majesty's Spanish Flock,* and Banks, *Sheep and Wool Correspondence.*
15. Stewart, *Sketches of the Character,* appendix, lvi. See also 172, 214.
16. Ibid., appendix, xlix.
17. Sinclair, *Address to the Society;* Sinclair, *Memoirs of the Life and Works,* 1:219–28.
18. Chambers, *Biographical Dictionary of Eminent Scotsmen,* 5:526. Most entries are available online at http://www.electricscotland.com/history/other/sinclair_john.htm (accessed 18 June 2009).
19. Sinclair, *Statistical Account of Scotland,* available at http://edina.ac.uk/stat-acc-scot/ (accessed 20 June 2009).
20. Sinclair, *Memoirs of the Life and Works,* 2:46.
21. Foucault, *Society Must Be Defended,* 240, 245, 246.
22. Ibid., 252.
23. See Gallagher and Greenblatt, *Practicing New Historicism,* 111–35.
24. Sinclair, *Memoirs of the Life and Works,* 2:46.
25. Wolfe, "Exposures," 8.
26. Derrida, "Animal That Therefore I Am," 396.
27. Burns, "To a Mouse," in *Complete Works,* 210.
28. Ibid., 106.

29. A more complete look at Burns's animals is beyond the scope of this chapter, but see also "The Twa Herds; Or, The Holy Tulyie," in which Burns writes, "An' get the brutes the power themsel's / To choose their herds." See also "The Inventory," addressed to Mr. Aitken of Ayr, surveyor of taxes for the district, and "The Death and Dying Words of Poor Mailie, the Author's Only Pet Yowe."

CHAPTER 5

1. "It is evident that the polis belongs to the class of things that exist by nature, and that man is by nature an animal intended to live in a polis. He who is without a polis, by reason of his own nature and not of some accident, is either a poor sort of being, or a being higher than man." Aristotle *Politics* 1.2.5. See also Guichet, *Usages politiques de l'animalité,* and Duchet, *Anthropologie et histoire.*
2. The term "animality" has received wide application within the field of animal studies. See Derrida, *Animal That Therefore I Am,* and a recent commentary by Lawlor, *This Is Not Sufficient.*—Trans.
3. See Dhombres and Dhombres, *Lazare Carnot.*
4. See Lacépède, Cuvier, and Geoffroy, *La ménagerie du Muséum d'histoire naturelle.*
5. See Blanckaert, "J.-J. Virey, observateur de l'homme" and "La perfectibilité sous condition?"
6. See Corsi, *Lamarck.*
7. See Chappey, *Société des observateurs.*
8. This gap in the literature is addressed by Paula Young Lee's forthcoming study, "Aristocrats and Other Animals."
9. The Consulate was the government of France between the fall of the republican government in Napoleon's coup of 18 Brumaire 1799 and the beginning of Napoleon's empire in 1804.—Trans.
10. See Osborne, "Applied Natural History and Utilitarian Ideals," and Chappey, *Société des observateurs.*
11. Jacques-Henri Bernardin de Saint Pierre to the Convention, MS 2528 (138), 21 January 1793, Muséum national d'histoire naturelle, Paris.
12. See Laissus and Petter, *Animaux du muséum.*
13. The date is from the French republican calendar, adopted by the revolutionaries in 1793 and used for about twelve years. Beginning at the autumn equinox, the calendar had twelve months of thirty days each, which were given new names based on nature.—Trans.
14. Papers of Geoffroy Saint-Hilaire, MS 2717 (2), Muséum national d'histoire naturelle, Paris.
15. Convention Nationale, *Rapport fait au nom du Comité d'instruction publique,* 21 Frimaire, Year III (11 December 1794).
16. Before the Revolution, Lacépède wrote two treatises, *Essai sur l'électricité* (1781) and *Physique générale et particulaire* (1782–84), which gained him the friendship of the comte de Buffon, who in 1785 appointed him subdemonstrator

in the Jardin du Roi and proposed that he continue Buffon's *Histoire naturelle.* This continuation was published under the titles *Histoire des quadrupèdes, ovipares et des serpents* (2 vols., 1788–89) and *Histoire naturelle des reptiles* (1789). When the Jardin du Roi was reorganized as the Jardin des Plantes, Lacépède was appointed to the chair allocated to the study of reptiles and fishes.

17. Lacépède, "Lettre relative aux établissements publics."
18. See Toscan, *L'ami de la nature.*
19. See Deleuze, *Histoire et description du Muséum royal.*
20. For example, the *Annales du Muséum national d'histoire naturelle.*
21. Geoffroy Saint-Hilaire, "Mémoire sur les orangs outangs." See also Barsanti, "L'orang-outang déclassé."
22. Audebert, *Histoire naturelle des singes.*
23. Audebert based this reference on the anthropological writings of Buffon, who called the Chacrelas an albino people of the island of Java.—Trans.
24. Cf. Audebert, *Histoire naturelle des singes,* following Camper, *Dissertation sur les variétés naturelles.* For additional context, see Blanckaert, "La classification des races."
25. Virey, *Histoire naturelle du genre humain.*
26. Corsi, *Lamarck,* 21–35.
27. Lamétherie, *Vues physiologiques* (unpaginated).
28. See Tinland, "Limites de l'animalité."
29. Lamétherie, *De l'homme,* xxii, xxxvi. In the foreword, Lamétherie expresses gratitude to Franz Joseph Gall, Samuel Thomas von Sömmering, and Johann Kaspar Lavater for their contribution to human knowledge.
30. Staum, "The Facial Angle," in *Labeling People,* 23–48.
31. Lamétherie, *De l'homme,* 401.
32. Ibid., 5–13.
33. Ibid., 65.
34. [Lamétherie], *Principes de la philosophie naturelle* (author's name handwritten on the copy in the Bibliothèque nationale), 204, 234. In fact, the author imagines a fictional case where a race of animals could become strong enough to defeat man and in turn reduce him to slavery, a recurrent fear in the Western imagination of the inversion of the forms of domination, evident from the prints of the sixteenth century to *Planet of the Apes* (235).
35. See Mannucci, "'Malheur aux faibles!'"
36. [Lamétherie], *Principes de la philosophie naturelle,* 246.
37. Ibid., 263.
38. Ibid., 250–57. The case of China and of its overpopulation particularly worried the author, who saw only three solutions: either the loan of land by other, less populated counties, the possibility of settlement in Tartary, or placing "limits on the multiplication" of the population.
39. See Bénot, *Démence coloniale sous Napoléon,* and Dorigny, *Rétablissement de l'esclavage.*
40. See Thomson, "Grégoire et l'unité de l'espèce humaine."
41. I plan to address this concept of infra-citizenship (*infra-citoyenneté*) in my next book on animals.

See Cohen, *Semi-Citizenship in Democratic Politics.*

42. Lamétherie, *De l'homme,* lviii. Elsewhere, Lamétherie compared the Hottentots negatively to Newton, the Hottentots' society to that of Louis XIV or Pericles. ["Laïs" may refer to the hetaerae, or courtesans, of ancient Greece, enticing young women renowned for their beauty and intelligence. The most celebrated of them was the mistress of Alcibiades, who was later killed by the women of Thessaly out of jealousy of her beauty. "Hottentot" is the derogatory name given to the Khoikhoi people, or bushmen, of southern Africa.—Trans.]
43. Ibid., 14.
44. Ibid., 92.
45. Ibid., 255.
46. Declamations against "brutes within the Republic," "brutes within the circle of citizenship," and "brutes among the respectable [*honnêtes gens*] citizens" were advanced by conservative proponents of order, fearful of the lingering and, in their view, corrupting presence in French society of former *sans-culottes* from the popular classes: working-class people who were strong supporters of the Revolution, commonly known by their long trousers, in contrast to the *culottes,* or knee-length pants, worn by the Old Regime aristocracy. The latter, deemed supporters of the Terror, were also commonly decried as drinkers of blood.—Trans.
47. On Salaville's concern with republican morals, their formation, regeneration, and maintenance, see Serna, "1799, le retour du refoulé." Well known in republican circles from the beginning of the Revolution, Salaville served as the secretary to the prominent early revolutionary the comte de Mirabeau. He was also the translator of radical seventeenth-century English thinkers.
48. *Procès verbaux des séances générales pour les années IX–XII de la république française* (1800–1804), Years IX and X (1800–1802), registre 3 A4, Archives de l'Institut national, Paris. One of the last, but hardly the least important, measures of the Thermidorean Convention was, by the law of 3 Brumaire, Year IV (25 October 1795), the founding of the National Institute, thereby incarnating the republican temple of science. It was divided into three classes: physical and mathematical science, moral and political science, and literature and the fine arts, each of them subdivided into several sections. This learned assembly reunited the most eminent scholars of the republic. Each of its classes represented a place where discussions occurred on the progress and possibilities for transformation and concrete amelioration of the economy, society, and the sciences. The National Institute found itself at the summit of the pyramid of the project for regeneration by education and the diffusion of knowledge, carried out by the republican political class. Following the Terror, it tried to achieve institutional stability by placing the project of education and

intellectual formation at the heart of civic construction, as the first guarantee of social order and morality in the republic. Cf. Staum, "Class of Moral and Political Sciences."

49. For the context of the English representatives' sensitivity to the animal question in the early nineteenth century, see Singer, *Animal Liberation,* 212ff.
50. See Sepinwall, *Abbé Grégoire and the French Revolution.*
51. This fact surely explains why this competition remained for so long unknown, as it was not deposited in the archives of the class of moral science. Its identification number (*côte*) is now in the Archives de l'Académie des inscriptions et belles-lettres, dossier 1 H8. The latter is, in effect, the academy, as restored in 1816, that gathered the holdings of the former institutions, thus assuring the continuity of the patrimony of the archives of the competitions.
52. See *Procès verbaux des séances générales pour les années IX–XII de la république française* (1800–1804), 2 Germinal, Year XII (23 March 1804), registre 3 A4, Archives de l'Institut national, Paris. A critical edition of the twenty-six *mémoires* was prepared under my supervision. The sociologist Vincent Pelosse is the first researcher to have undertaken an analysis of this competition. See his "Imaginaire social et protection de l'animal."
53. Only one other manuscript was published: Granchamp's *Essai philosophique.*
54. In 1754 Condillac published a treatise on animals at the end of his *Treatise on Sensations.* An attack on the systems of Descartes and Buffon, it presents a radically different hypothesis concerning animal nature. For the eighteenth-century followers of the seventeenth-century philosopher of reason René Descartes, the animal is the result of a mechanical organization, which reduces it to a living machine devoid of all sensitivity (*sensibilité*). In his *Histoire naturelle,* Buffon develops another hypothesis concerning the difference between animals and men. Animals, according to Buffon, are simple creatures, limited only to their instinct, devoid of all forms of thought or intelligence, but not lacking *sensibilité.* Men, on the contrary, are complex creatures endowed with a double nature, one animal, the other resolutely spiritual, which renders them intelligent, perfectible, capable of adaptation and constant improvement. In contrast, for Condillac, who opposed the ideas of both Descartes and Buffon, the only difference between man and animal is one of degree (*du plus où moins*). The two beings are tied together by a similar nature. However, having much more limited needs than man, the animal's senses are changed into habits, whereas man invents constantly, imagines relentlessly. The debate between these positions characterized French science and philosophy for more than a century. On the influence of Condillac in the construction

of the republican science of the Directory and its influence on the *idéologues,* on the second class of the institute, and on the members of the moral and political section, in the years 1797–1805, see the excellent account by Malino, *Jew in the French Revolution.*

55. "If the alleged ill-treatment of animals makes man, in effect, atrocious and barbarous to his fellow creatures, hunters, butchers, and all those who, by condition, mistreat or kill animals, the surgeons who dissect them alive, the physiologists who perform experiments on them that they cannot make on man, in order to try to discover the mechanism and the secret of the game of life, could they preserve in themselves the slightest trace of humanity? The habit of torturing and slaughtering animals, do they not distort it to the point that they make a game of tormenting and killing people? This is, however, what never happens. Everyone may have known hunters and butchers as human, as compassionate, and sometimes more so, than those who accuse them not to be." Salaville, *De l'homme et des animaux,* 40–41. With regard to butchers, Salaville asks, "have you searched the offices of the criminal clerks of all nations to see if this class yields more murderers, more evildoers than the others? Is this not one of those allegations that cannot withstand the test of scrutiny, but persist because no one dares to undergo this test?" (41).

56. Fontenay, *Silence des bêtes.*

57. Salaville, *De l'homme et des animaux,* 6.

58. On the persistence of Descartes's thought in the eighteenth century, see Le Ru, "Réception occasionaliste de Descartes."

59. Salaville, *De l'homme et des animaux,* 16–17.

60. Salaville writes, "How to introduce some generosity into habits, some magnanimity into ideas still under the influence of the constant principle of barbarism and inhumanity? In this respect, the position of civilized peoples will be rather less favorable than that of savage peoples, because they only kill animals. Unlike us, they don't have the art of reducing them to servitude; and in the series of ill treatments toward men, slavery is certainly the worst of all because it is the most corrupting, that is to say, it is the one which most alters in our souls the principles of justice and humanity." Ibid., 48. See also Dorlin, *Matrice de la race.*

61. "One who seeks to lead social man to happiness must thus teach him not to let himself be dominated by disordered feelings. He will show him that true happiness lies in knowing how to enjoy in peace the benefits of nature"; "all that we have just said concerning the rich man in society whose fortune provides abundantly for what he desires, without being obliged to resort to work, it is him to which we will principally occupy ourselves in this work. But for the one who is attached daily to the fields [*glèbe*]

or to his workshop, what would we be able to say to him? His task is fixed. He is condemned to work." Lamétherie, *De l'homme* 13–14. Lamétherie wrote elsewhere, "Thus the Negro differs less from the orang-outang than the latter differs from the mandrill [a large West African baboon]. I will go further; I say that such a Negro differs more from the physique of Apollo or Antinous [originally a member of Emperor Hadrian's entourage, commemorated in medals and statues and known for his extraordinary beauty] than from the orang-outang." Lamétherie, *Considérations sur les êtres organisés,* 55.

CHAPTER 6

1. Lockwood, *Natural History,* 237.
2. Defoe, *Tour Through the Whole Island of Great Britain,* 291.
3. De Quincey, *Literary Reminiscences,* 310–11.
4. Ritvo, *Animal Estate,* 284–89.
5. Darwin, *On the Origin of Species,* 73–74.
6. Williams, *Keywords,* 184.
7. Cronon, "Trouble with Wilderness," 69.
8. Tankerville and Hindmarsh, "On the Wild Cattle of Chillingham Park," 611.
9. For an extended discussion of the history of this debate, see Ritvo, "Race, Breed, and Myths of Origin."
10. For summaries of recent research, see Hall and Clutton-Brock, *Two Hundred Years of British Farm Livestock;* Hall, "White Herd of Chillingham," 112–19; and Hall, "Running Wild," 12–15, 46–49.
11. Bennet, "Chillingham Cattle," 22.
12. Ritvo, *Animal Estate,* 232–42.
13. Koerner, *Linnaeus,* 87–88.
14. Linnaeus, *Systema naturae,* 24.
15. See, for example, the illustrations in Thomas Bewick's popular *General History of Quadrupeds,* first published in 1790.
16. Peel, *Zoological Gardens of Europe,* 205–6; "In Memory of Consul," pamphlet in the Belle Vue Collection, Chetham's Library, Manchester, UK.
17. Quoted in Wilson, *Signs and Portents,* 56–67.
18. White, *Account of the Regular Gradation in Man,* 34; Blumenbach, *Anthropological Treatises,* 73.
19. White, *Account of the Regular Gradation in Man,* 34.
20. Blumenbach, *Anthropological Treatises,* 80–81.
21. Bondeson and Miles, "Julia Pastrana, the Nondescript," 199.
22. Reported in *Nature* (12 May 1882), cited in Howard, *Victorian Grotesque,* 56–57.
23. Bailey and Culley, *General View of the Agriculture of Northumberland,* 245.
24. Dickinson, "On the Farming of Cumberland," 264.
25. Ian Herbert, "Foot and Mouth Crisis: Cumbria," *Independent,* 27 March 2001, 5.
26. Darwin, *On the Origin of Species,* 50.
27. Ritvo, *Platypus and the Mermaid,* 92–95.
28. Clutton-Brock, *Natural History of Domesticated Mammals,* 194–97.

CHAPTER 7

1. Russell, *Prince of Wales' Tour,* 252–54.
2. See Rothfels, "Killing Elephants."
3. *Appleton's Journal,* "Prince of Wales in India," 1.
4. Ibid., 10.
5. *Ballou's Monthly Magazine,* "Prince Alfred and His Elephant Hunt," 400.
6. See Haraway, "Teddy Bear Patriarchy," 49.
7. Ibid., 25.
8. Alvey, "Cinema as Taxidermy," 41, 42.
9. Ibid., 42.
10. Akeley, *In Brightest Africa,* 54.
11. Alvey, "Cinema as Taxidermy," 28.
12. Susan McHugh, e-mail message to author, 4 June 2009.
13. Akeley, *In Brightest Africa,* quotations on 175–77, 27.
14. "Bronx Zoo Elephant Chained for 2 Years: Visitors Stirred to Pity for Gunda, Bound in His Cage After Attacking Keeper," *New York Times,* 23 June 1914.
15. H. V. L., letter to the editor, *New York Times,* 26 June 1914.
16. William T. Hornaday, letter to the editor, *New York Times,* 27 June 1914.
17. "Bullet Ends Gunda, Bronx Zoo Elephant: Dr. Hornaday Ordered Execution Because Gunda Reverted to Murderous Traits," *New York Times,* 23 June 1915.
18. Alvey, "Cinema as Taxidermy," 42.
19. William T. Hornaday, letter to the editor, *New York Times,* 7 February 1913.
20. I am particularly indebted here to Diana Donald's reading of Sir Edwin Landseer's *A Random Shot* (1848) in her *Picturing Animals in Britain,* 300–305. For Donald, the work demands that we see the emotions and intellectual sentiments embedded in trophy hunting as much more complex than pure bloodlust.
21. Akeley, *In Brightest Africa,* 114–18.

CHAPTER 8

1. "Recreational Cod Fishermen Fined for Illegal Fishing," *Telegram* (St. John's, Newfoundland), 9 April 2010, http://www.thetelegram.com/Business/Natural-resources/2010-04-09/article-1453916/Recreational-cod-fishermen-fined-for-illegal-fishing/1 (accessed 15 January 2011).
2. UN FAO, Fisheries and Aquaculture Department, *State of World Fisheries and Aquaculture, 2008;* World Bank, "Sunken Billions."
3. Pauly, Watson, and Alder, "Global Trends in World Fisheries," 8–10.
4. Whitty, "Fate of the Ocean," 6.
5. Pauly and Watson, "Counting Down the Last Fish," 45.
6. See Fisher, "Hungry Oceans."
7. Belton, "High Seas Drifters," 34.
8. Myers and Worm, "Rapid Worldwide Depletion of Predatory Fish Communities," 281.
9. Macintosh, "Some Cod Populations at Historic Lows." For the official pronouncement, see Committee on the Status of Endangered Wildlife in Canada, *COSEWIC Status Assessments.*
10. Simpson, "Cod in Newfoundland: Already Seen That Drama," *Globe and Mail* (Toronto), 13 May 2010.

11. "To remain internationally competitive, Canadian producers must sustain the relentless pursuit of technological and management improvements that allow Canada to gain stature in world aquaculture. The capability to produce and market desired products at internationally competitive prices is paramount to sustained development." This is the Canadian government's recommendation for its aquaculture development strategy. See Canada, Department of Fisheries and Oceans, *Federal Aquaculture Development Strategy.*
12. For an overview of the contrast between subsistence and industrialized fishing, see UN FAO, Fisheries and Aquaculture Department, "Small-Scale Fisheries Around the World" and "Fisheries and Aquaculture Report #911."
13. The family-based inshore cod fishery that prevailed roughly between the 1790s and 1930s was partly a reflex of merchant capitalism, as argued persuasively in Sider, *Between History and Tomorrow.* In the centuries before the nineteenth, there were few "planters" or settlers on the maritime coast. Most fishermen, hired as servants for wages, were brought over from Spain, Portugal, France, and England to work on the North Atlantic fisheries between the melting ice of spring and the encroaching fog and storms of fall, which made the transatlantic passage perilous.
14. Perret, *La géographie de terre-neuve,* 185–86, quoted in Innis, *Cod Fisheries,* 2–3. For a fuller account of the modes and manner of cod fishing off the Grand Banks, see Kurlansky, *Cod,* 1–17.
15. See Innis, *Cod Fisheries,* 4.
16. Anspach, *History of Newfoundland,* 354–55, quoted in Sider, *Between History and Tomorrow,* 72.
17. Cadigan, in "Moral Economy of the Commons," documents the activities of fishermen and their families unrelated to fishing by which they deepened their competency and their independence of fish merchants.
18. Morgan, *Bristol and the Atlantic Trade,* 55–88.
19. Sider, *Between History and Tomorrow,* 25–29, 79–83.
20. Sider insists that the truck system always worked to the disadvantage of the fisherman, whereas Cadigan provides evidence of paternalistic merchants carrying fishermen for many years, knowing full well that without a winter supply many would face destitution, if not starvation.
21. See Agamben, *Means Without Ends,* for the distinction between "form-of-life" and form of life. Agamben writes of "form-of-life," "a life that cannot be separated from its form is a life for which what is at stake in its way of living is living itself" (4).
22. Blake, "International Fishery," 208.
23. "Aquaculture is the culture of aquatic organisms, including fish, molluscs, crustaceans and aquatic plants. Culture implies some form of intervention in the rearing process to enhance production, such as regular stocking, feeding, protection from predators, etc. Culture also implies

individual or corporate ownership of the stock being cultivated." Canada, Department of Fisheries and Oceans, *Federal Aquaculture Development Strategy.*

24. "Pronounced aversion toward measurements and numbers . . . is admissible when it is a manner of gaining a quick overview about the manifold varieties of organic forms, and is pardonable when the pleasure of the composing artist in the beauty and variety of forms and in his [God's] fanciful conceptions is greater than the sense for exploration of the analytical scholar; but this aversion toward measurement and numbers, which at times is heightened into contempt, is incomprehensible, inadmissible, and unpardonable when the scholar demands that his labours be regarded as a contribution to the knowledge of the true laws of nature." Friedrich Heincke, quoted in Sinclair and Solemdal, "Development of 'Population Thinking,'" 195.
25. The history of the MSY construct and its political versus scientific virtues, in particular for the cold war era in the United States, are detailed in Finley, "Political History of Maximum Sustained Yield," and Smith, *Scaling Fisheries.*
26. Finlayson, *Fishing for Truth,* 219–27.
27. Gordon's "Economic Approach to the Optimum Utilization" is an exemplary instance of bio-economic models of fish and fisherman.
28. Sinclair, "State Encloses the Commons," 157–77.
29. Sider, *Between History and Tomorrow,* 33. See also 29–35 for the sociopolitical consequences of the collapse of the cod fisheries off the Grand Banks. For a moving historical account of the transformation of the cod fisheries from abundance to scarcity, see chapter 3 of Nicholls, *Paradise Found.* See also Bavington, *Managed Annihilation,* 1–2.
30. For the intrinsic vagueness of system-theory constructs in the fisheries and the likely lack of their operationalization, see Garcia et al., "Ecosystem Approach to Fisheries." To locate the position of ecosystem models on the vector of scientific models of cod fisheries, see Caddy, "Fisheries Management in the Twenty-First Century," and Caddy and Cochrane, "Review of Fisheries Management."
31. The domestication of wild nature for global markets is aptly illustrated in a recent trade publication: "You cannot develop new markets without a consistent supply of product, and steady, predictable year-round production is the goal for every new species that is being brought into aquaculture. *It is a question of getting animals to spawn when you want them to rather than when nature dictates.* Scientists and industry are finding ways to alter the timing of these natural events to their advantage." Lockett, "Aquaculture," 56 (emphasis added).
32. Ruzzante et al., "Bay-Scale Population Structure in Coastal Atlantic Cod," 442. See also Royal

Society of Canada, *Elements of Precaution.*

33. Naylor et al., "Effects of Aquaculture on World Fish Supplies," 1020, and Pauly et al., "Fishing Down and Farming Up the Food Web," 25.
34. Belton, "High Seas Drifters," 34.
35. Conley, "Environmental Concerns," 48; Marshall, *Fishy Business,* 110.
36. Lockwood, "Who Is Capturing Aquaculture's Values?" 32.
37. Simpson, "Cod in Newfoundland."
38. For a glimpse into the many attempts by fishing people to protest the introduction of threshold-breaching fishing gear, see in particular chapter 10 in Roberts, *Unnatural History of the Sea.* For examples of such dissent and protest specific to the cod fisheries of Newfoundland and Labrador, consult the magisterial work of Innis, *Cod Fisheries.* In *Hope and Deception in Conception Bay,* Cadigan details the vexed relationships between merchants and fishermen in these parts. Cadigan, in "Moral Economy of the Commons," also describes the wrenching impact on native peoples of Europeans who settled the coastlines, and discusses the fishermen's arguments against capital-intensive and productivity-enhancing technologies. For the repeated failure of techno-science in constructing the cod fisheries, see also Cadigan, "Failed Proposals for Fisheries Management" and "Moral Economy of Retrenchment and Regeneration."
39. The pivotal argument of this chapter concerning natural thresholds is derived from Illich, *Tools for Conviviality.* We also rely on this work for the implicit distinction between tools that can be "used" and those that must be "operated." See also Illich's *Shadow Work* for elaborations of this concept applied to "science" and "economy." For the distinction between common sense and scientific rationality, see Arendt, "Conquest of Space and the Stature of Man."
40. To grasp the extent of the political repercussions resulting from the de facto criminalization of fishermen, see "Grimes Says Ban on Cod Fishing 'Dead Wrong,'" CTV News, 7 May 2003, http://montreal.ctv.ca/servlet/an/local/CTVNews/20030507/grimes_cod_nfld_030507?hub=OttawaHome (accessed 13 November 2004).
41. See, for example, Power, "Women Processing Workers."
42. "By allowing professional fish harvesters to make their own business decisions and be accountable for the consequences . . . [DFO hopes] to spawn a new and positive fisheries management culture and usher in a new era of public-private sector co-operation in Canada's fisheries." Canada, Department of Fisheries and Oceans, "Atlantic Policy Review," sec. 5.3. For the state-sponsored transformation of fish hunters into professional managers of biomass, see also Professional Fish Harvesters Certification Board, Newfoundland and Labrador, *Frequently Asked*

Questions, http://www.pfhcb.com/ (accessed 12 October 2004).

CHAPTER 9

1. *Anekdotomania* is Spoerri's collection of autobiographical anecdotes about his oeuvre up till 2001. Unless otherwise noted, all translations in this chapter are my own.
2. In the visual arts, assemblage as an artistic process is typically related to the experiments of such artists as Marcel Duchamp, Pablo Picasso, and, especially, Jean Dubuffet's early 1950s series of collages—of butterfly wings titled *assemblages d'empreintes,* where objects not originally intended as art materials are introduced in a new context. See Seitz, *Art of Assemblage,* and Geiger, *Art of Assemblage.*
3. Spoerri, *Anekdotomania,* 114, 118.
4. Ibid., 237. On Le Brun's physiognomic studies, see Montagu, *Expression of the Passions.* See also Gareau, *Charles Le Brun.*
5. Jennifer Montagu underscores the innovative features of Le Brun's work especially in relation to Giovanni Battista della Porta's physiognomic study, also mentioned by Spoerri (*Anekdotomania,* 238). However, Montagu plays down the role of the physiognomic in Le Brun's work on expression. With regard to human and animal resemblances, she explains Le Brun's plates both as pioneering (indeed, as almost visionary, and anticipatory of Darwin's studies) and as indebted to the contemporaneous faith in reason, especially Descartes's philosophy. On these grounds, she views Le Brun's resemblances as a great improvement over della Porta's "crude" drawings. Montagu, *Expression of the Passions,* 20–21.
6. See Foucault, *Order of Things.*
7. Spoerri, *Anekdotomania,* 238.
8. Lee connects representations of animals with the fierce debates concerning perspective in the seventeenth century. She argues that at the core of the debate was the view of animals, for the live animal face as it is found in nature—i.e., in motion and in time—is distinct from an architectural landscape and resists being interpreted through the laws of mathematical perspective. The foundation of rational human vision, in short, was controversial from the beginning, and the law of perspective could be said to stand on unstable ground. See Lee, "Taming the Two-Eyed Beast" (courtesy of the author), and Montagu, *Expression of the Passions,* 23.
9. Spoerri, *Anekdotomania,* 97.
10. On the great chain of being, see the eponymous study by Lovejoy, especially chapter 8, 227–41.
11. Montagu, *Expression of the Passions,* 66–67, 71–73.
12. Ibid., 18.
13. Ibid., 27.
14. Or else the human is the depth/death of the animal, namely, its vanishing point.
15. Deleuze and Guattari, *Kafka,* 28, 29, 59.
16. See Baker, "'You Kill Things to Look at Them.'" Speaking of the effects

of animal corpses in art, Baker quotes Elaine Scarry (81–83). The quotation perfectly renders the effect of the *détrompe l'oeil* as it applies to *Carnival.*

17. Deleuze and Guattari, *Kafka,* 28, 29, 59.
18. See Agamben, *Open,* 91–92.
19. See Mitchell's analysis of the duck-rabbit, which these ducks recall, in *Picture Theory,* 51, 53–57.
20. Agamben, *Open,* 26–27.
21. "'Becoming-animal' as a form of 'becoming-minoritarian' is an important instance of what Deleuze and Guattari term micropolitics, a politics of desire that at crucial points departs from the agenda of the 'politics of identity' and pursues a more radical trajectory than that of representation or emancipation." Urpeth, "Animal Becomings," 211n6.
22. Spoerri's account of the geographical difference between the Siberian and Alaskan coats redoubles—parodies—Beuys's statements about East and West, his idea of "Eurasia." Beuys indeed mentions the migration of the wolf from Siberia to America in this context. On the wolf-coyote, see Beuys, *Energy Plan for the Western Man,* 213.
23. Ibid., 142.
24. See Deleuze and Guattari, *Thousand Plateaus,* 35, 43.
25. Spoerri had once illustrated the expression "it's raining cats and dogs" in one of his "word traps," or *Wortfallen.* See Spoerri, *Anekdotomania,* 140–44. Here, a tool is freed of its use and becomes a substantive of its fixity.
26. Bakhtin, *Rabelais and His World,* 39.
27. Ibid., 39, 40.
28. Spoerri, "'La messa in scena degli oggetti,'" 44.
29. Baker, "'You Kill Things to Look at Them,'" 74.
30. Foucault, *Order of Things,* xv.
31. The French title of Foucault's work, translated into English as *The Order of Things.*
32. On the avant-garde and carnival, Alfred Jarry's *Ubu Roi,* and Luis Buñuel's *Exterminating Angel,* see Stam, *Subversive Pleasures,* 94–111.

bibliography

Adams, Carol J. *The Sexual Politics of Meat: A Feminist-Vegetarian Critical Theory.* Cambridge: Polity Press, 1990.

Agamben, Giorgio. *Means Without Ends: Notes on Politics.* Translated by Vincenzo Binetti and Cesare Casarino. Minneapolis: University of Minnesota Press, 2000.

———. *The Open: Man and Animal.* Translated by Kevin Atell. Stanford: Stanford University Press, 2004.

Akeley, Carl. *In Brightest Africa.* New York: Doubleday, 1923.

Akeley, Delia J. *Jungle Portraits.* New York: Macmillan, 1930.

Alvey, Mark. "The Cinema as Taxidermy: Carl Akeley and the Preservative Obsession." *Framework: The Journal of Cinema and Media* 48, no. 1 (2007): 23–45.

Anderson, Virginia DeJohn. *Creatures of Empire: How Domestic Animals Transformed Early America.* Oxford: Oxford University Press, 2004.

Animal Studies Group, ed. *Killing Animals.* Urbana: University of Illinois Press, 2006.

Anonymous. *A newe mery and wittie Comedie or Enterlude, newely imprinted, treating vpon the Historie of Iacob and Esau.* London: Henry Binneman, 1568.

Anspach, Lewis. *A History of Newfoundland.* London: T. and J. Allman, 1819.

Appleton's Journal: A Monthly Miscellany of Popular Literature. "The Prince of Wales in India," July 1877, 1–15.

Arendt, Hannah. "The Conquest of Space and the Stature of Man." In Arendt, *Between Past and Future,* 265–80. New York: Penguin Books, 1993.

Armitage, David. "John Locke, Carolina, and the *Two Treatises of Government.*" *Political Theory* 32 (2004): 602–27.

Audebert, Jean-Baptiste. *Histoire naturelle des singes et des makis.* Paris: Desray, Year VIII [1800].

Babington, Gervase. *Certaine Plaine, Briefe, and Comfortable Notes vpon Euerie Chapter of Genesis.* London: Thomas Charde, 1592.

Bachelor [Batchelor], Thomas. *Village Scenes, the Progress of Agriculture, and Other Poems.* London: Parnassian Press, 1804.

Bailey, John, and George Culley. *General View of the Agriculture of Northumberland, Cumberland, and Westmorland* (1805). Newcastle, UK: Frank Graham, 1972.

Baker, Steve. *Picturing the Beast: Animals, Identity, and Representation.* Urbana: University of Illinois Press, 2001.

———. *The Postmodern Animal.* London: Reaktion Books, 2000.

———. "'You Kill Things to Look at Them': Animal Death in Contemporary Art." In *Killing Animals,* ed. Animal Studies Group, 69–98. Urbana: University of Illinois Press, 2006.

Bakhtin, Mikhail M. *Rabelais and His World.* Translated by Hélène Iswolsky. Bloomington: Indiana University Press, 1984.

Baldassari, Anne. *Picasso: Life with Dora Maar; Love and War, 1935–1945.* Paris: Flammarion, 2006.

Ballou's Monthly Magazine. "Prince Alfred and His Elephant Hunt," 1 April 1868, 400.

Banks, Joseph. *The Sheep and Wool Correspondence of Sir Joseph Banks, 1781–1820.* Edited by Harold Burnell Carter. London: Library Council of NSW in association with the British Museum (Natural History), 1979.

Barkan, Leonard, Bradin Cormack, and Sean Keilen, eds. *The Forms of Renaissance Thought: New Essays in Literature and Culture.* Basingstoke, UK: Palgrave, 2009.

Barsanti, Giulio. "L'orang-outang déclassé: Histoire du premier signe à hauteur d'homme (1780–1801) et ébauche d'une théorie de la circularité des sources." *Bulletin et Mémoires de la Société d'Anthropologie de Paris* 1, nos. 3–4 (1989): 67–104.

Barthes, Roland. *Elements of Semiology.* Translated by Annette Lavers and Colin Smith. New York: Hill and Wang, 1977.

———. *Mythologies.* Translated by Annette Lavers. New York: Farrar, Straus and Giroux, 1972.

Bavington, Dean. *Managed Annihilation: An Unnatural History of the Newfoundland Cod Collapse.* Vancouver: University of British Columbia Press, 2010.

Bell, Millicent. "Naked Lear." *Raritan* 23, no. 4 (2004): 57–70.

Belton, Ben. "High Seas Drifters: The Reckless Deep-Sea Gold Rush That Could Turn the North American Continental Shelf into One Giant Factory-Style Fish Farm." *Ecologist* 34, no. 6 (2004): 32–38.

Bennet, Ian. "Chillingham Cattle." *Ark* 18 (1991): 22.

Bénot, Yves. *La démence coloniale sous Napoléon.* Paris: La Découverte, 2006.

Bernasconi, Robert, and Anika Maaza Mann. "The Contradictions of Racism: Locke, Slavery, and the *Two Treatises.*" In *Race and Modern Philosophy,* ed. Andrew Valls, 89–107. Ithaca: Cornell University Press, 2005.

Berry, Edward. *Shakespeare and the Hunt: A Cultural and Social Study.* Cambridge: Cambridge University Press, 2001.

Beuys, Joseph. *Energy Plan for the Western Man: Joseph Beuys in America; Writings by and Interviews with the Artist.* Compiled by Carin Kuoni. New York: Four Walls Eight Windows, 1990.

Blake, Raymond. "The International Fishery off Canada's East Coast in the Twentieth Century." In *How Deep Is the Ocean: Historical Essays on Canada's Atlantic Fishery,* ed. James Candow and Carol Corbin, 207–22. Sydney, Nova Scotia: Cape Breton University Press, 1997.

Blanckaert, Claude. "La classification des races au XVIIIe siècle: L'anthropologie naturaliste entre méthode et anti-méthode." In "L'invention et la représentation des races au XVIIIe siècle," ed. Isabelle Baudino, special issue, *Lumières* 14 (2009): 13–41.

———. "J.-J. Virey, observateur de l'homme (1800–1825)." In *Julien-Joseph Virey, naturaliste et anthropologue,* ed. Claude Benichou and Claude Blanckaert, 97–182. Paris: Vrin, 1998.

———. "La perfectibilité sous condition? Education d'espèce, flexibilité d'organisation et échelle d'aptitude morale en anthropologie, 1750–1820." In *L'homme perfectible,* ed. Bertrand Binoche, 114–44. Seyssel: Champ Vallon, 2004.

Bloomfield, Robert. *The Farmer's Boy.* London: Vernor and Hood, 1800.

Blumenbach, Johann Friedrich. *The Anthropological Treatises. . . .* Edited and translated by Thomas Bendyshe. London: Longman, Green, Longman, Roberts and Green / Anthropological Society, 1865.

Bodry-Sanders, Penelope. *Carl Akeley: Africa's Collector, Africa's Savior.* New York: Paragon, 1991.

Bondeson, Jan, and A. E. W. Miles. "Julia Pastrana, the Nondescript: An Example of Congenital Generalized Hypertrichosis Terminalis with Gingival Hyperplasia." *American Journal of Medical Genetics* 47 (1993): 198–212.

Bosman, Anston. "Shakespeare in Leather." In *The Forms of Renaissance Thought: New Essays in Literature and Culture,* ed. Leonard Barkan, Bradin Cormack, and Sean Keilen, 225–45. Basingstoke, UK: Palgrave, 2009.

Bowden, Peter J. *The Wool Trade in Tudor and Stuart England.* London: Frank Cass, 1962.

Brown, Bill. "How to Do Things with Things (a Toy Story)." *Critical Inquiry* 24, no. 4 (1998): 935–64.

———. "Thing Theory." *Critical Inquiry* 28, no. 1 (2001): 1–22.

Bruster, Douglas. *Shakespeare and the Question of Culture: Early Modern Literature and the Cultural Turn.* New York: Palgrave, 2003.

Buffon, Georges-Louis Leclerc, comte de. *Histoire naturelle, générale et particulière, avec la description du Cabinet du roi.* 44 vols. Paris: Imprimerie Royale, 1749–1804.

———. "The 'Initial Discourse' to Buffon's 'Histoire Naturelle':

The First Complete English Translation." Translated by John Lyon. *Journal of the History of Biology* 9, no. 1 (1976): 133–81.

———. *Natural History, General and Particular . . . Translated into English . . . With . . . Notes and Observations by the Translator.* Trans. William Smellie. 8 vols. London: A. Strahan and T. Cadell, 1781.

Burns, Robert. *The Complete Works of Robert Burns.* Edited by Allan Cunningham. Boston: Phillips, Sampson, 1855.

Caddy, John. "Fisheries Management in the Twenty-First Century: Will New Paradigms Apply?" *Reviews in Fish Biology and Fisheries* 9 (1999): 1–43.

Caddy, John, and Ken Cochrane. "A Review of Fisheries Management Past, Present, and Some Future Perspectives for the Third Millennium." *Oceans and Coastal Management* 44 (2001): 653–82.

Cadigan, Sean T. "Failed Proposals for Fisheries Management and Conservation in Newfoundland, 1855–1880." In *Fishing Places, Fishing People,* ed. Dianne Newell and Rosemary E. Ommer, 147–69. Toronto: University of Toronto Press, 1999.

———. *Hope and Deception in Conception Bay: Merchant-Settler Relations in Newfoundland, 1785–1855.* Toronto: University of Toronto Press, 1995.

———. "The Moral Economy of the Commons: Ecology and Equity in the Newfoundland Cod Fishery, 1815–1855." *Labour/Le Travail* 43 (1999): 9–42.

———. "The Moral Economy of Retrenchment and Regeneration in the History of Rural Newfoundland." In *Retrenchment and Regeneration in Rural Newfoundland,* ed. Reginald Byron, 14–42. Toronto: University of Toronto Press, 2003.

Calarco, Matthew, and Peter Atterton, eds. *Animal Philosophy: Essential Readings in Continental Thought.* London: Continuum Press, 2004.

Calvin, John. *A Commentarie of Iohn Caluine, upon the first booke of Moses called Genesis.* London: John Harrison and George Bishop, 1578.

Camper, Pierre. *Dissertation sur les variétés naturelles qui caractérisent la physionomie des hommes des différents climats et des différents âges.* Translated by Hendrick J. Jansen. Paris: Francart, 1792.

Canada, Government of. Department of Fisheries and Oceans. "Atlantic Policy Review." http://www.dfo-mpo.gc.ca/afpr-rppa/home_e.htm.

———. *Federal Aquaculture Development Strategy.* http://www.dfo-mpo.gc.ca/aquaculture/ref/FADS-SFDA-eng.htm#tphp (accessed 10 May 2010).

Carruthers, Peter. *The Animals Issue: Moral Theory in Action.*

Cambridge: Cambridge University Press, 1992.

Carter, Harold Burnell. *His Majesty's Spanish Flock: Sir Joseph Banks and the Merinos of George III of England.* London: Angus and Robertson, 1964.

Césaire, Aimé. *Discourse on Colonialism.* Translated by Joan Pinkham. New York: Monthly Review Press, 1972.

Chambers, Robert. *Biographical Dictionary of Eminent Scotsmen, with a Supplemental Volume.* Edited by Thomas Thomson. 5 vols. Edinburgh: Blackie and Son, 1856.

Chappey, Jean-Luc. *La Société des observateurs de l'homme (1799–1804), des anthropologues au temps de Bonaparte.* Paris: Société des Études Robespierristes, 2002.

———. "Usages et enjeux politiques d'une métaphorisation de l'espace savant en Révolution: 'L'encyclopédie vivante' de la république thermidorienne à l'empire." *Politix* 48 (1999): 37–70.

Clarkson, Leslie A. "The Organization of the English Leather Industry in the Late Sixteenth and Seventeenth Centuries." *Economic History Review* 13, no. 2 (1960): 245–56.

Clutton-Brock, Juliet. *Domesticated Animals from Early Times.* London: British Museum (Natural History), 1981.

———. *A Natural History of Domesticated Mammals.* London: British Museum (Natural History), 1987.

Cohen, Elizabeth. *Semi-Citizenship in Democratic Politics.* Cambridge: Cambridge University Press, 2009.

Committee on the Status of Endangered Wildlife in Canada. *COSEWIC Status Assessments,* May 2003. http://www.cosewic.gc.ca/htmlDocuments/Detailed_Species_Assessment_e.htm (accessed 4 May 2003).

Conley, David. "Environmental Concerns: The Anti-Salmon Farming Lobby in BC." *Aquaculture Magazine,* April 1998, 36–51.

Connell, Philip. *Romanticism, Economics, and the Question of "Culture."* Oxford: Oxford University Press, 2001.

Connor, Steven. "Thinking Things." *Textual Practice* 24, no. 1 (2010): 1–20.

Convention Nationale. *Rapport fait au nom du Comité d'instruction publique et des finances, sur le Muséum d'histoire naturelle par Thibaudeau, à la séance du 21 frimaire l'an III.* 11 December 1794. Paris: Imprimerie Nationale, 1794.

Corbin, Alain. *The Foul and the Fragrant: Odor and the French Social Imagination* Cambridge: Harvard University Press, 1988.

Corsi, Pietro. *Lamarck: Genèse et enjeux du transformisme, 1770–1830.* Paris: CNRS, 2001.

Cousin, John William. *A Short Biographical Dictionary of*

English Literature. London: J. M. Dent and Sons, 1910.

Cox, Nancy, and Karin Dannehl, eds. *Dictionary of Traded Goods and Commodities, 1550–1820.* British History Online, 2007, http://www.british-history.ac.uk/source.aspx?pubid=739 (accessed 16 February 2010).

Cox, Neil, and Deborah Povey. *A Picasso Bestiary.* London: Academy Editions, 1995.

Cronon, William. "The Trouble with Wilderness; or, Getting Back to the Wrong Nature." In Cronon, *Uncommon Ground: Rethinking the Human Place in Nature,* 69–90. New York: W. W. Norton, 1996.

Cummings, Brian. "Pliny's Literate Elephant and the Idea of Animal Language in Renaissance Thought." In *Renaissance Beasts: Of Animals, Humans, and Other Wonderful Creatures,* ed. Erica Fudge, 164–85. Urbana: University of Illinois Press, 2004.

Curth, Louise Hill. *Care of the Brute Beast: A Social and Cultural History of Veterinary Medicine in Early Modern England.* Leiden: Brill, 2010.

Dallas, R. C. *A History of the Maroons from Their Origin to the Establishment of Their Chief Tribe at Sierra Leone.* 2 vols. London: T. M. Longman and O. Rees, 1803.

Dance, S. Peter. *The Art of Natural History.* New York: Arch Cape, 1990.

Dannenfeldt, Karl H. "Europe Discovers Civet Cats and Civet." *Journal of the History of Biology* 18, no. 3 (1985): 403–31.

Danner, Mark. "Abu Ghraib: The Hidden Story." *New York Review of Books,* 7 October 2004. http://www.nybooks.com/articles/17430 (accessed 20 April 2009).

Darwin, Charles. *On the Origin of Species.* 1859. Facsimile edition. Cambridge: Harvard University Press, 1964.

Daston, Lorraine, and Katharine Park. *Wonders and the Order of Nature, 1150–1750.* New York: Zone Books, 1998.

Defoe, Daniel. *A Tour Through the Whole Island of Great Britain, 1724–1726.* Edited by P. N. Furband and W. R. Owens. New Haven: Yale University Press, 1991.

Deleuze, Gilles. "Body, Meat, and Spirit: Becoming-Animal." In Deleuze, *Francis Bacon: The Logic of Sensation,* trans. Daniel W. Smith, 19–24. Minneapolis: University of Minnesota Press, 2003.

Deleuze, Gilles, and Félix Guattari. *Kafka: Toward a Minor Literature.* Translated by Dana Polan. Minneapolis: University of Minnesota Press, 1986.

———. *A Thousand Plateaus: Capitalism and Schizophrenia.* Translated by Brian Massumi. Minneapolis: University of Minnesota Press, 1987.

Deleuze, Joseph Philippe François. *Histoire et description du*

Muséum royal d'histoire naturelle, ouvrage rédigé d'après les ordres de l'administration du muséum. Paris: Royer, 1823.

Démoris, René. "Chardin, la machine et l'oiseau." *Revue des Sciences Humaines* 186–87 (April–October 1982): 293–306.

———. "Inside/Interiors: Chardin's Images of the Family." *Art History* 28, no. 4 (2005): 442–67.

De Quincey, Thomas. *Literary Reminiscences from the Autobiography of an English Opium Eater.* Vol. 3 of *The Works of Thomas De Quincey,* 12 vols. Boston: Houghton Mifflin, 1876.

Derrida, Jacques. *The Animal That Therefore I Am.* Edited by Marie-Louise Mallet. Translated by David Wills. New York: Fordham University Press, 2008.

———. "The Animal That Therefore I Am (More to Follow)." *Critical Inquiry* 28, no. 2 (2002): 369–418.

Descartes, René. "René Descartes' Sixth Reply to Thomas Hobbes's Objections." 1641. Translated by George MacDonald Ross. http://www.philosophy.leeds.ac.uk/GMR/hmp/texts/modern/hobbes/objections/objects.html (accessed 20 July 2011).

Despret, Vinciane. "Sheep Do Have Opinions." In *Making Things Public: Atmospheres of Democracy,* ed. Bruno Latour and Peter Weibel, 360–68. Cambridge: MIT Press, 2005.

Dhombres, Jean, and Nicole Dhombres. *Lazare Carnot.* Paris: Fayard, 1997.

Dickinson, William. "On the Farming of Cumberland." *Journal of the Royal Agricultural Society of England* 13 (1852): 207–300.

Dion, Mark, Lisa G. Corrin, Miwon Kwon, and Norman Bryson. *Mark Dion* London: Phaidon Press, 1997.

Dion, Mark, Lawrence Weschler, D. Graham Burnett, and Rebecca Solnit. *The Marvelous Museum: Orphans, Curiosities, and Treasures; A Mark Dion Project.* San Francisco: Chronicle Books, 2010.

Donald, Diana. *Picturing Animals in Britain.* New Haven: Yale University Press, 2007.

Donne, John. "To Sir Edward Herbert, at Juliers." In *John Donne: The Complete English Poems,* ed. A. J. Smith, 218. London: Penguin Books, 1985.

Dorigny, Marcel. *Le rétablissement de l'esclavage dans les colonies françaises: 1802 aux origines de Haïti.* Paris: Maisonneuve et Larose, 2003.

Dorlin, Elsa. *La matrice de la race, généalogie sexuelle et coloniale de la nation française.* Paris: La Découverte, 2006.

Duchet, Michèle. *Anthropologie et histoire au siècle des lumières.* Paris: Flammarion, 1971.

[Edwards, Bryan, ed.] *The Proceedings of the Governor and Assembly of Jamaica, in Regard to the Maroon*

Negroes. London: Stockdale, 1796.

Farr, James. "Locke, Natural Law, and New World Slavery." In *Locke's Political Liberty: Readings and Misreadings,* ed. Christophe Miqueu and Mason Chamie, 165–88. Oxford: Voltaire Foundation, 2009.

Ferry, Luc. *The New Ecological Order.* Translated by Carol Volk. Chicago: University of Chicago Press, 1995.

Finlayson, Alan Christopher. *Fishing for Truth: A Sociological Analysis of Northern Cod Stock Assessments from 1977–1990.* St. John's, Newfoundland: ISER Books, 1994.

Finley, Carmel. "A Political History of Maximum Sustained Yield, 1945–1955." In *Oceans Past: Management Insights from the History of Marine Animal Populations,* ed. David Starkey, Paol Holm, and Michaela Barnard, 189–214. London: Earthscan, 2008.

Fisher, Emily. "Hungry Oceans: What Happens When All the Prey Are Gone?" *Oceana,* 2 March 2009. http://na.oceana.org/en/blog/2009/03/hungry-oceans (accessed 5 July 2009).

Fontenay, Élisabeth de. *Le silence des bêtes: La philosophie à l'épreuve de l'animalité.* Paris: Fayard, 1998.

Foster, Hal. *The Return of the Real: The Avant-Garde at the Turn of the Century.* Cambridge: MIT Press, 1996.

Foucault, Michel. *The Order of Things: An Archaeology of the Human Sciences.* New York: Vintage Books, 1973.

———. *Society Must Be Defended: Lectures at the Collège de France, 1975–1976.* Translated by David Macey. New York: Picador, 2003.

Freedberg, David. *The Eye of the Lynx: Galileo, His Friends, and the Beginnings of Modern Natural History.* Chicago: University of Chicago Press, 2002.

Freud, Sigmund. *The Wolf Man and Other Cases.* Translated by Louise Adey Huish. London: Penguin Classics, 2003.

Fudge, Erica. *Perceiving Animals: Humans and Beasts in Early Modern English Culture.* New York: St. Martin's Press, 2000.

———, ed. *Renaissance Beasts: Of Animals, Humans, and Other Wonderful Creatures.* Urbana: University of Illinois Press, 2004.

Gallagher, Catharine, and Stephen Greenblatt. *Practicing New Historicism.* Chicago: University of Chicago Press, 2001.

Garcia, S. M., A. Zerbi, C. Aliaume, T. Do Chi, and G. Lasserre. "The Ecosystem Approach to Fisheries: Issues, Terminology, Principles, Institutional Foundations, Implementation, and Outlook." FAO Fisheries Technical Paper 443. Rome: UN Food and Agriculture Organization, 2003.

Gardiner, Samuel Rawlson, ed. *Documents Relating to the Proceedings Against William Prynne in 1634 and 1637.* London: Camden Society, 1877.

Gareau, Michel. *Charles Le Brun: First Painter to King Louis XIV.* New York: Harry N. Abrams, 1992.

Garvey, Eleanor M. *The Artist and the Book, 1860–1960: In Western Europe and the United States; Exhibition Held at the Museum of Fine Arts, Boston, May 4–July 16, 1961.* Exh. cat. Boston: Museum of Fine Arts, 1962.

Geiger, Stephan. *The Art of Assemblage, the Museum of Modern Art, 1961: Die neue Realität der Kunst in den frühen sechziger Jahren.* Munich: Verla Silke Schreiber, 2008.

Geoffroy Saint-Hilaire, Étienne. "Extrait d'un mémoire sur les orangs outangs lu à la société philomatique par le citoyen Geoffroy, professeur de zoologie au Muséum d'histoire naturelle." *Bulletin de Sciences par la Société Philomatique* 2 (July 1797): 25–26.

———. "Mémoire sur les orangs outangs." *Journal de Physique* 46 (Year VI [1798]): 185–91.

Gessner, Conrad. *Icones animalium quadrupedum viviparorum et oviparorum quae in historia animalium.* Zurich: Christoffel Froschover, 1560.

Goeppert, Sebastian, Herma Corinna Goeppert, Patrick Cramer, and Pablo Picasso. *Pablo Picasso, the Illustrated Books: Catalogue raisonné.* Geneva: Patrick Cramer, 1983.

Gombrich, Ernst. *Art and Illusion: A Study in the Psychology of Pictorial Representation.* New York: Phaidon, 2002.

Gordon, H. S. "An Economic Approach to the Optimum Utilization of Fishery Resources." *Journal of the Fisheries Research Board of Canada* 10 (1993): 442–57.

Granchamp, Joseph Louis. *Essai philosophique sur la question de savoir jusqu'à quel point les traitements barbares exercés sur les animaux, intéressent-ils la morale publique; et conviendrait-il de faire des lois à cet régard?* Paris: Faine Jeune, Year XII [1804].

Guerrini, Anita. *Experimenting with Humans and Animals: From Galen to Animal Rights.* Baltimore: Johns Hopkins University Press, 2003.

———. "The 'Virtual Menagerie': The Histoire des Animaux Project." *Configurations: A Journal of Literature, Science, and Technology* 14, nos. 1–2 (2000): 29–41.

Guichet, Jean Luc, ed. *Usages politiques de l'animalité.* Paris: L'Harmattan, 2008.

Hadfield, Andrew. "Plants in *King Lear.*" *Notes and Queries* 57, no. 3 (2010): 385–86.

Hall, Stephen J. G. "Running Wild." *Ark* 16 (1989): 12–15, 46–49.

———. "The White Herd of Chillingham." *Journal of the Royal Agricultural Society of England* 150 (1989): 112–19.

Hall, Stephen J. G., and Juliet Clutton-Brock. *Two Hundred Years of British Farm Livestock.* London: British Museum (Natural History), 1988.

Haraway, Donna Jeanne. *The Companion Species Manifesto: Dogs, People, and Significant Otherness.* Chicago: Prickly Paradigm Press, 2003.

———. "Teddy Bear Patriarchy: Taxidermy in the Garden of Eden, New York City, 1908–36." *Social Text* 11 (Winter 1984–85): 19–64.

———. *When Species Meet.* Minneapolis: University of Minnesota Press, 2008.

Harris, Jonathan Gil. *Untimely Matter in the Time of Shakespeare.* Philadelphia: University of Pennsylvania Press, 2009.

Harrison, Peter. "Reading Vital Signs: Animals and the Experimental Philosophy." In *Renaissance Beasts: Of Animals, Humans, and Other Wonderful Creatures,* ed. Erica Fudge, 186–207. Urbana: University of Illinois Press, 2004.

Henninger-Voss, Mary J., ed. *Animals in Human Histories: The Mirror of Nature and Culture.* Rochester: University of Rochester Press, 2002.

Herbert, Ian. "Foot and Mouth Crisis: Cumbria." *Independent,* 27 March 2001, 5.

Horodisch, Abraham. *Picasso as a Book Artist.* London: Faber and Faber, 1962.

Howard, Martin. *Victorian Grotesque: An Illustrated Excursion into Medical Curiosities, Freaks, and Abnormalities—Principally of the Victorian Age.* London: Jupiter Books, 1977.

Illich, Ivan. *Shadow Work.* London: Marion Boyars, 1985.

———. *Tools for Conviviality.* Berkeley, Calif.: Heyday Books, 1973.

Innis, Harold. *The Cod Fisheries: The History of an International Economy.* Toronto: University of Toronto Press, 2004.

Introna, Lucas D. "Ethics and the Speaking of Things." *Theory, Culture, and Society* 26, no. 4 (2009): 398–419.

Ivison, Duncan. "Locke, Liberalism, and Empire." In *The Philosophy of John Locke: New Perspectives,* ed. Peter R. Arnstey, 86–105. London: Routledge, 2003.

Johnson, Sara E. "'Give Them Blacks to Eat': Waging Inter-American Wars of Terror and Torture." *American Quarterly* 61, no. 1 (2009): 65–92.

Johnson, Thomas. *Cornucopiae, or Diuers Secrets Wherein is Contained the Rare Secrets in Man, Beasts, Foules, Fishes, Trees, Plantes, Stones and Such Like.* London: W. Barley, 1595.

Jones, Ann Rosalind, and Peter Stallybrass. *Renaissance Clothing and the Materials of Memory.* Cambridge: Cambridge University Press, 2000.

Kaloff, Linda. *Looking at Animals in Human History.* New York: Reaktion Books, 2007.

Kamber, André, ed. *Petit lexique sentimental autour de Daniel Spoerri.* Paris: Musée des Beaux-Arts and Centre Georges Pompidou, 1990.

Kay, Sarah. "Legible Skins: Animals and the Ethics of Medieval Reading." *Postmedieval: A Journal of Medieval Cultural Studies* 2, no. 1 (2011): 13–32.

Keefer, Michael H. "Accommodation and Synecdoche: Calvin's God in King Lear." *Shakespeare Studies* 20 (1988): 147–68.

Kete, Kathleen. *The Beast in the Boudoir: Petkeeping in Nineteenth-Century Paris.* Berkeley and Los Angeles: University of California Press, 1994.

Knoppers, Laura Lunger, and Joan B. Landes. *Monstrous Bodies/Political Monstrosities in Early Modern Europe.* Ithaca: Cornell University Press, 2004.

Koerner, Lisbet. *Linnaeus: Nature and Nation.* Cambridge: Harvard University Press, 1999.

Kopytoff, Igor. "The Cultural Biography of Things: Commoditization as Process." In *The Social Life of Things: Commodities in Cultural Perspective,* ed. Arjun Appadurai, 64–93. Cambridge: Cambridge University Press, 1986.

Kurlansky, Mark. *Cod: A Biography of a Fish That Changed the World.* London: Vintage, 1999.

La Capra, Dominick. *History and Its Limits: Human, Animal, Violence.* Ithaca: Cornell University Press, 2009.

Lacépède, Bernard Germain Étienne de Laville-sur-Illon, de. "Lettre relative aux établissements publics destinés à renfermer des animaux vivants, et connus sous le nom de ménageries." *Décade Philosophique, Litteraire et Politique* (20 Frimaire, Year IV [11 December 1795]): 449–62.

Lacépède, Bernard Germain Étienne de Laville-sur-Illon de, Georges Cuvier, and Étienne Geoffroy Saint-Hilaire. *La ménagerie du Muséum d'histoire naturelle ou description et histoire des animaux qui y vivent et qui y ont vécu.* Paris: Miger et Renouard, Year XII [1804].

Laissus, Yves, and Jean-Jacques Petter. *Les animaux du muséum, 1793–1993.* Paris: Imprimerie Nationale, 1993.

Lamétherie, Jean-Claude de. *Considérations sur les êtres organisés.* Paris: H. L. Perroneau, Year XIII [1805].

———. *De l'homme considéré moralement, de ses mœurs et de celles des animaux.* Paris: Maradan Libraire sur Pavée Saint André des Arts, 1802.

———. *Principes de la philosophie naturelle dans lesquels on cherche à déterminer les degrés de certitude et de probabilité des connaissances humaines.* Geneva, 1787.

———. *Vues physiologiques sur l'organisation animale et végétale.* Amsterdam: Didot, 1780.

Landry, Donna. "Georgic Ecology." In *Robert Bloomfield: Lyric, Class, and the Romantic Canon,* ed. Simon White, John Goodridge, and Bridget Keegan, 253–68. Lewisburg: Bucknell University Press, 2006.

Latour, Bruno. "Morality and Technology: The End of the Means." Translated by Couze Venn. *Theory, Culture, and Society* 19, nos. 5–6 (2002): 247–60.

Latour, Bruno, and Peter Weibel, eds. *Making Things Public: Atmospheres of Democracy.* Cambridge: MIT Press, 2005.

Lavin, Irving. "Picasso's Bull(s): Art History in Reverse." *Art in America* 81, no. 3 (1993): 76–123.

Law, John. "Notes on the Theory of the Actor-Network: Ordering, Strategy, and Heterogeneity." *Systems Practice* 5, no. 4 (1992): 379–93.

Lawlor, Leonard. *This Is Not Sufficient: An Essay on Animality and Human Nature in Derrida.* New York: Columbia University Press, 2007.

Laws of Jamaica, 1681–1758. St. Jago de la Vega, Jamaica: Alexander Aikman, 1796.

Lebovics, Herman. *Imperialism and the Corruption of Democracies.* Durham: Duke University Press, 2006.

Lee, Paula Young. "Aristocrats and Other Animals: Political Death and Animal Life at the Revolutionary Ménagerie in Paris." Manuscript.

———, ed. *Meat, Modernity, and the Rise of the Slaughterhouse.* Durham: University of New Hampshire Press, 2008.

———. "Taming the Two-Eyed Beast: Images, Animals, and Objects in the Seventeenth-Century French Academies." Paper presented at "Beyond Mimesis and Nominalism: Representation in Art and Science," University College of London, 22–23 June 2006. http://philsci-archive.pitt.edu/2799/.

Le Roy, Charles-Georges. *Lettres sur les animaux.* Edited by Elisabeth Anderson. Oxford: Voltaire Foundation, 1994.

Le Ru, Véronique. "La réception occasionaliste de Descartes: Des malebranchistes à l'Encyclopédie." *Recherches sur Diderot et l'Encyclopédie* 38 (2005): 191–202.

Levy, Thomas, ed. *Daniel Spoerri: Coincidence as Master.* Bielefeld: Kerber, 2004.

Liebman, Elizabeth Amy. "Painting Natures: Buffon and the Art of the 'Histoire naturelle.'" PhD diss., University of Chicago, 2003.

Linnaeus, Carolus. *Systema naturae: Regnum animale.* 1735. London: British Museum (Natural History), 1956.

Lippincott, Louise, and Andreas Blühm. *Fierce Friends: Artists and Animals, 1750–1900.* London: Merrell, 2005.

Lippit, Akira Mizuta. *Electric Animal: Toward a Rhetoric of Wildlife.* Minneapolis: University of Minnesota Press, 2000.

Locke, John. *Political Essays.* Edited by Mark Goldie. Cambridge: Cambridge University Press, 1997.

———. *Two Treatises of Government.* Edited by Peter Laslett. 3rd ed. Cambridge: Cambridge University Press, 1988.

Lockett, John. "Aquaculture: What's on the Menu?" *Atlantic Business Magazine,* June–July 2001, 52–62.

Lockwood, Edward. *Natural History, Sport, and Travel.* London: W. H. Allen, 1878.

Lockwood, Glen. "Who Is Capturing Aquaculture's Values?" *Aquaculture Magazine,* January–February 1999, 30–33.

Lovejoy, Arthur O. *The Great Chain of Being: A Study of the History of an Idea.* Cambridge: Harvard University Press, 1936. Reprint, New Brunswick, N.J.: Transaction Publishers, 2009.

Loveland, Jeff. *Rhetoric and Natural History: Buffon in Polemical and Literary Context.* Oxford: Voltaire Foundation, 2001.

Macintosh, Zoë. "Some Cod Populations at Historic Lows." *Live Science,* 13 May 2010. http://www.livescience.com/environment/cod-fish-endangered-100513.html (accessed 15 May 2010).

Malino, Frances. *A Jew in the French Revolution: The Life of Zalkind Hourwitz.* Oxford: Blackwell, 1996.

Mannucci, Erica. "'Malheur aux faibles!' Condamnation de l'oppression des animaux." *Dix-huitième Siècle* 28 (1996): 353–66.

Margolies, Eleanor. "Vagueness Gridlocked: A Map of the Smells of New York." In *The Smell Culture Reader,* ed. Jim Drobnick, 107–17. New York: Berg, 2006

Marshall, Dale. *Fishy Business: The Economics of Salmon Farming in B.C.* Vancouver: Canadian Centre for Policy Alternatives, 2003. http://www.livingoceans.org/documents/fishy%20business.pdf (accessed 2 February 2004).

Matarasso, Henri, and Pierre-André Benoit. *Bibliographie des livres illustrés par Pablo Picasso: Oeuvres graphiques, 1905–1956.* Nice: Galerie H. Matarasso, 1956.

Mazzio, Carla. "The Senses Divided: Organs, Objects, and Media in Early Modern England." In *The Empire of the Senses: The Sensual Culture Reader,* ed. David Howes, 85–105. New York: Berg, 2005.

McEarthron, Scott, ed. *Nineteenth-Century English Labouring Class Poets, 1800–1900.* Vol. 1. London: Pickering and Chatto, 2006.

McShane, Clay, and Joel A. Tarr. *The Horse in the City: Living Machines in the Nineteenth Century.* Baltimore: Johns Hopkins University Press, 2007.

Michael, Mark A. "Locke's *Second Treatise* and the Literature of Colonization." *Interpretation* 25 (1998): 407–27.

Mitchell, William J. Thomas. *Picture Theory.* Chicago: University of Chicago Press, 1994.

Montagu, Jennifer. *The Expression of the Passions: The Origin and Influence of Charles Le Brun's "Conférence sur l'expression générale et particulière."* New Haven: Yale University Press, 1994.

Montaigne, Michel de. "An Apologie of *Raymond Sebond.*" In *Montaigne's Essays,* trans. John Florio. Renascence Editions: An Online Repository of Works Printed in English Between the Years 1477 and 1799. https://scholarsbank.uoregon.edu/xmlui/bitstream/handle/1794/766/montaigne.pdf?sequence=1 (accessed 19 May 2010).

Morgan, Kenneth. *Bristol and the Atlantic Trade in the Eighteenth Century.* Cambridge: Cambridge University Press, 1993.

Mourlot, Fernand. *Picasso Lithographs.* Translated by J. Didry. Boston: Boston Book and Art, 1970.

Muséum national d'histoire naturelle. *Annales du Muséum national d'histoire naturelle.* 20 vols. Paris: Frères Levrault, 1802–13.

Myers, Ram, and Boris Worm. "Rapid Worldwide Depletion of Predatory Fish Communities." *Nature* 423 (2003): 280–83.

Naylor, R., R. Goldberg, J. Primaversa, N. Kautsky, M. Beveridge, J. Clay, K. Folke, J. Lubchenco, H. Mooney, and M. Troell. "Effects of Aquaculture on World Fish Supplies." *Nature* 405 (2000): 1017–24.

Neeson, Jeanette M. *Commoners: Common Right, Enclosure, and Social Change in England, 1700–1820.* Cambridge: Cambridge University Press, 1993.

Neill, Michael. "'Amphitheaters in the Body': Playing with Hands on the Shakespearian Stage." *Shakespeare Survey* 48 (1995): 25–50.

Nicholls, Steve. *Paradise Found: Nature in America at the Time of Discovery.* Chicago: University of Chicago Press, 2009.

Novero, Cecilia. *Antidiets of the Avant-Garde: From Futurist Cooking to Eat Art.* Minneapolis: University of Minnesota Press, 2010.

O'Neal, John C. *The Authority of Experience: Sensationist Theory in the French Enlightenment.* University Park: Pennsylvania State University Press, 1996.

Osborne, Michael A. "Applied Natural History and Utilitarian Ideals: 'Jacobin Science' at the Muséum d'histoire naturelle, 1789–1870." In *Re-Creating Authority in Revolutionary France,* ed. Bryan T. Ragan and Elisabeth A. Williams, 124–43. New Brunswick: Rutgers University Press, 1992.

Pauly, D., P. Tyedmers, R. Froese, and L. Y. Liu. "Fishing Down and Farming Up the Food Web." *Conservation Biology in Practice* 2, no. 4 (2001): 25.

Pauly, D., and R. Watson. "Counting the Last Fish." *Scientific American*, July 2003, 42–47.

Pauly, D., R. Watson, and J. Alder. "Global Trends in World Fisheries: Impacts on Marine Ecosystems and Food Security." *Philosophical Transactions of the Royal Society B: Biological Sciences* 360, no. 1453 (2005): 5–12.

Peel, C. V. A. *The Zoological Gardens of Europe: Their History and Chief Features.* London: F. E. Robinson, 1903.

Pelosse, Vincent. "Des amis des bêtes de l'an X au législateur de 1850." *L'Homme* 21, no. 4 (1981): 5–33, and 22, no. 1 (1982): 35–51.

Perrault, Claude. *Mémoires pour servir à l'histoire naturelle des animaux.* Published in two parts. Paris: Imprimerie Royale, 1671, 1676.

Perret, Robert. *La géographie de terre-nueve.* Paris: E. Guilmoto, 1913.

Philo, Chris, and Chris Wilbert. "Animal Spaces, Beastly Places: An Introduction." In *Animal Spaces, Beastly Places: New Geographies of Human-Animal Relations*, ed. Chris Philo and Chris Wilbert, 1–34. London: Routledge, 2000.

Picasso, Pablo. *Histoire naturelle: Picasso eaux-fortes originales pour les textes de Buffon* [Picasso's Original Etchings for Buffon's Text]. Paris: Martin Fabiani, 1942.

Picasso, Pablo, and Georges-Louis Leclerc, comte de Buffon. *40 dessins en marge du Buffon.* Paris: Jonquières, 1957.

Picasso, Pablo, Steven A. Nash, Robert Rosenblum, and Brigitte Baer. *Picasso and the War Years, 1937–1945.* Exh. cat. New York: Thames and Hudson, 1998.

Potts, Alex. "Natural Order and the Call of the Wild: The Politics of Animal Picturing." *Oxford Art Journal* 13, no. 1 (1990): 12–33.

Power, Nicole. "Women Processing Workers as Knowledgeable Resource Users: Connecting Gender, Local Knowledge, and Development in the Newfoundland Fishery." In *Finding Our Sea Legs: Linking Fishery People and Their Knowledge with Science and Management*, ed. Barbara Neis and Lawrence F. Felt, 189–204. St. John's, Newfoundland: ISER Books, 2000.

Prynne, William. *Histrio-Mastix: The Players Scourge, or Actors Tragedie.* London: Michael Sparke, 1633.

Pyne, William Henry. *Microcosm, or A Picturesque Delineation of the Arts, Agriculture, and Manufactures of Great Britain in a Series of Above a Thousand Groups of Small Figures for the Embellishment of Landscape.* Aquatinted by J. Hill, with an introduction by A. E. Santaniello and commentary by C. Gray. 2 vols. London: William Miller, 1806. Reprint, New York: Benjamin Blom, 1971.

Rainsford, Marcus. *An Historical Account of the Black Empire of Hayti.* London: Cundee, 1805.

Reill, Peter Hanns. *Vitalizing Nature in the Enlightenment.* Berkeley and Los Angeles: University of California Press, 2005.

Ritvo, Harriet. *The Animal Estate: The English and Other Creatures in the Victorian Age.* Cambridge: Harvard University Press, 1987.

———. *The Platypus and the Mermaid, and Other Figments of the Classifying Imagination.* Cambridge: Harvard University Press, 1997.

———."Race, Breed, and Myths of Origin: Chillingham Cattle as Ancient Britons." *Representations* 39 (Summer 1992): 1–22.

Robbins, Louise E. *Elephant Slaves and Pampered Parrots: Exotic Animals in Eighteenth-Century Paris.* Baltimore: Johns Hopkins University Press, 2002.

Roberts, Callum. *The Unnatural History of the Sea.* London: Island Press, 2007.

Roger, Jacques. *Buffon: A Life in Natural History.* Translated by Sarah Lucille Bonnefoi. Ithaca: Cornell University Press, 1997.

———. *Les sciences de la vie dans la pensée française du XVIIIe siècle.* Paris: Armand Colin, 1971.

Rothfels, Nigel. "Killing Elephants: Pathos and Prestige in the Nineteenth Century." In *Victorian Animal Dreams: Representations of Animals in Victorian Literature and Culture,* ed. Deborah Denenholz Morse and Martin Danahay, 53–63. Burlington, Vt.: Ashgate, 2007.

———, ed. *Representing Animals.* Bloomington: University of Indiana Press, 2002.

Royal Society of Canada. *Elements of Precaution: Recommendations for the Regulation of Food Biotechnology in Canada; An Expert Panel Report on the Future of Food Biotechnology.* Ottawa: Royal Society of Canada, 2001.

Russell, William Howard. *The Prince of Wales' Tour: A Diary in India; with Some Account of the Visits of His Royal Highness to the Courts of Greece, Egypt, Spain, and Portugal.* New York: Lovell, Adam, Wesson, 1877.

Ruzzante, D., J. Wroblewski, C. Taggart, R. Smedbol, D. Cook, and S. Goddard. "Bay-Scale Population Structure in Coastal Atlantic Cod in Labrador and Newfoundland." *Canadian Journal of Fisheries Biology* 56 (2000): 431–47.

Salaville, Jean-Baptiste. *De l'homme et des animaux.* Paris: Derteville, Year XIII [1805].

Sanchez, Ricardo S. "Memorandum for Combined Joint Task Force Seven." 14 September 2003. http://www.cup.columbia.edu/media/3738/jaffer-blog.pdf (accessed 23 July 2011).

Saussure, Ferdinand de. *Writings in General Linguistics.* Edited by Simon Bouquet and Rudolf Engler. Translated by Carol Sanders and Matthew Pires. Oxford: Oxford University Press, 2006.

Seitz, William C. *The Art of Assemblage.* Exh. cat. New York: Museum of Modern Art, 1961.

Senior, Matthew. "The Ménagerie and the Labyrinthe: Animals at Versailles, 1662–1792." In *Renaissance Beasts: Of Animals, Humans, and Other Wonderful Creatures,* ed. Erica Fudge, 208–32. Urbana: University of Illinois Press, 2004.

Sepinwall, Alyssa Goldstein. *The Abbé Grégoire and the French Revolution: The Making of Modern Universalism.* Berkeley and Los Angeles: University of California Press, 2005.

Serna, Pierre. "1799, le retour du refoulé ou l'histoire de la Révolution Anglaise à l'ordre du jour de la crise du Directoire." In *La Révolution, 1789–1871: Écriture d'une histoire immédiate,* ed. Philippe Bourdin, 241–58. Clermont-Ferrand: Presses Universitaires Blaise Pascal, 2008.

Shakespeare, William. *King Lear.* Ed. Kenneth Muir. London: Routledge, 1989.

———. *The Merchant of Venice.* In *The Oxford Shakespeare,* ed. Gary Taylor and Stanley Wells, 453–80. Oxford: Clarendon Press, 1988.

Shannon, Laurie. "Poor, Bare, Forked: Animal Sovereignty, Human Negative Exceptionalism, and the Natural History of King Lear." *Shakespeare Quarterly* 60, no. 2 (2009): 168–96.

Sider, Gerald. *Between History and Tomorrow: Making and Breaking Everyday Life in Newfoundland.* 2nd ed. Ontario: Broadview Press, 2003.

Simmons, A. John. *On the Edge of Anarchy: Locke, Consent, and the Limits of Society.* Princeton: Princeton University Press, 1993.

Sinclair, John. *Address to the Society for the Improvement of British Wool; Constituted at Edinburgh, on Monday January 31, 1791.* London: Printed for T. Cadell, 1791.

———. *Statistical Account of Scotland.* 22 vols. Edinburgh: W. Creech, 1791–99. Reprint, Edinburgh: University of Edinburgh Press; and Glasgow: University of Glasgow Press, 1999.

Sinclair, John, Jr. *Memoirs of the Life and Works of the Late Right Honorable Sir John Sinclair, Bart.* 2 vols. Edinburgh: William Blackwood and Sons; London: T. Cadell, 1837.

Sinclair, Michael, and Per Solemdal. "The Development of 'Population Thinking' in Fisheries Biology Between 1878 and 1930." *Aquatic Living Resources* 1 (1988): 189–213.

Sinclair, Peter R. "The State Encloses the Commons: Fisheries Management from the 200-Mile Limit to Factory Freezer Trawlers." In *A Question of Survival: The Fisheries and Newfoundland Society,* ed. Peter R. Sinclair, 157–77. St. John's,

Newfoundland: ISER Press, 1988.

Singer, Peter. *Animal Liberation: A New Ethics for Our Treatment of Animals.* New York: Discus, 1977.

Sloan, Philip R. "The Buffon-Linnaeus Controversy." *Isis* 67 (1976): 356–75.

———. "From Logical Universals to Historical Individuals: Buffon's Idea of Biological Species." In *Histoire de concept d'espèce dans les sciences de la vie,* ed. Scott Atran, 101–40. Paris: Fondation Singer-Polighac, 1987.

———. "The Gaze of Natural History." In *Inventing Human Science: Eighteenth-Century Domains,* ed. Christopher Fox, Roy Porter, and Robert Wokler, 112–51. Berkeley and Los Angeles: University of California Press, 1995.

Sloan, Philip R., and John Lyon, eds. and trans. *From Natural History to the History of Nature: Readings from Buffon and His Critics.* Notre Dame: University of Notre Dame Press, 1981.

Smith, Tim D. *Scaling Fisheries: The Science of Measuring the Effects of Fishing, 1855–1955.* Cambridge: Cambridge University Press, 1994.

Spoerri, Daniel. *Anekdotomania: Daniel Spoerri über Daniel Spoerri.* Basel: Hatje Cantz Verlag, 2001.

———. *Le carnaval des animaux, d'après Charles Le Brun: Physiognomonia, figures humaines comparées avec celles des animaux.* Exh. cat. Edited by Mark Gisbourne. Paris: Galerie Yvon Lambert, 1996.

———. *From Trap Pictures to Prillwitz Idols.* Edited by Thomas Levy, Barbara Räderscheidt, and Sandra Solimano. Translated by Barbara Baroni and Marina Pugliano. Milan: Silvana, 2010.

———. "'La messa in scena degli oggetti': Una conversazione con Daniel Spoerri." Interview by Sandro Parmiggiani. In *Daniel Spoerri: La messa in scena degli oggetti.* Milan: Skira, 2004.

———. *Werke: 1960–2001.* Bielefeld: Kerber, 2001.

Spoerri, Daniel, and Robert Filliou. *Anekdoten zu einer Topographie des Zufalls.* Hamburg: Nautilus Verlag, 1998. Translated by Robert Filliou, Emmett Williams, Dieter Roth, and Roland Topor as *An Anecdoted Topography of Chance* (London: Atlas, 1995).

Stallybrass, Peter. "Worn Worlds: Clothes and Identity on the Renaissance Stage." In *Subject and Object in Renaissance Culture,* ed. Margreta de Grazia, Maureen Quilligan, and Peter Stallybrass, 289–320. Cambridge: Cambridge University Press, 1996.

Stallybrass, Peter, and Ann Rosalind Jones. "Fetishizing the Glove in Renaissance Europe." *Critical Inquiry* 28, no. 1 (2001): 114–32.

Stam, Robert. *Subversive Pleasures: Bakhtin, Cultural Criticism, and*

Film. Baltimore: Johns Hopkins University Press, 1989.

Staum, Martin S. "The Class of Moral and Political Sciences, 1795–1803." *French Historical Studies* 11, no. 3 (1980): 371–97.

———. *Labeling People: French Scholars on Society, Race, and Empire, 1815–1848.* Montreal: McGill-Queens University Press, 2003.

Stewart, David. *Sketches of the Character, Manners, and Present State of the Highlanders of Scotland.* 3rd ed. Edinburgh: Longman, Rees, Orme, Brown, and Green, 1825.

Stewart, Susan. "Remembering the Senses." In *The Empire of the Senses: The Sensual Culture Reader,* ed. David Howes, 59–69. New York: Berg, 2005.

Stubbes, Philip. *The Anatomie of Abuses.* London: J. R. Jones, 1583.

Symmes, Marilyn F. "Illustrated Books at the Toledo Museum of Art." *Journal of Decorative and Propaganda Arts* 7 (1988): 52–71.

Tankerville, C. A. B., and L. Hindmarsh. "On the Wild Cattle of Chillingham Park." *Athenaeum* 565 (25 August 1838): 611–12.

Thomas, Keith. *Man and the Natural World: A History of the Modern Sensibility.* New York: Pantheon, 1983.

Thompson, Nato, ed. *Becoming Animal: Contemporary Art in the Animal Kingdom.* Exh. cat. Cambridge: MIT Press, 2005.

Thomson, Ann. "Grégoire et l'unité de l'espèce humaine." In *Grégoire et la cause des noirs, 1789–1831: Combats et projets,* ed. Yves Bénot and Marcel Dorigny, 11–23. Paris: Société Française d'Histoire d'Outre-mer, 2005.

Tinland, Frank. "Les limites de l'animalité et de l'humanité selon Buffon et leur pertinence pour l'anthropologie contemporaine." In *Buffon 88,* ed. Jean Gayon, 543–56. Paris: Vrin, 1992.

Topsell, Edward. *The Historie of Fovre-Footed Beastes.* London: William Jaggard, 1607.

Toscan, Georges. *L'ami de la nature, ou choix d'observation sur divers objets de la nature ou de l'art, suivi d'un catalogue de tous les animaux qui se trouvent actuellement dans la ménagerie.* Paris: Crapelet, Year VIII [1799].

United Nations. Food and Agriculture Organization, Fisheries and Aquaculture Department. "Fisheries and Aquaculture Report #911." Rome: UN FOA Fisheries and Aquaculture Department, 2008.

———. "Small-Scale Fisheries Around the World." http://www.fao.org/fishery/ssf/world/en (accessed 23 May 2010).

———. *State of World Fisheries and Aquaculture, 2008.* Rome: UN FOA Fisheries and Aquaculture Department, 2009.

United States Army. *Army Regulation 190-12, Military Working Dog Program.* Washington, D.C.: Department of the Army, 1993 and 2007. http://www.fas.org/irp/doddir/army/ar190-12.pdf.

———. *Army Regulation 190-14, Carrying of Firearms and Use of Force for Law Enforcement and Security Duties.* Washington, D.C.: Department of the Army, 1993. http://www.apd.army.mil/pdffiles/190_14.pdf.

Urpeth, James. "Animal Becomings." In *Animal Philosophy: Essential Readings in Continental Thought,* ed. Matthew Calarco and Peter Atterton, 101–10. London: Continuum, 2004.

Violand-Hobi, Heidi E. *Daniel Spoerri: Biographie und Werke.* Munich: Prestel, 1998.

Virey, Julien-Joseph. *Histoire naturelle du genre humain.* Paris: F. Dufart, Year IX [1800].

Weintraub Gallery. *Picasso: Histoire naturelle de Buffon; Thirty-One Cancelled Original Copper Plates.* Exh. cat. New York: Weintraub Gallery, 1971.

White, Charles. *An Account of the Regular Gradation in Man, and in Different Animals and Vegetables; and from the Former to the Latter.* London: C. Dilly, 1799.

Whitty, Julia. "The Fate of the Ocean." *Mother Jones,* March–April 2006, 1–12. http://motherjones.com/politics/2006/03/fate-ocean (accessed 7 August 2007).

Williams, Raymond. *Keywords: A Vocabulary of Culture and Society.* Oxford: Oxford University Press, 1976.

Wilson, Dudley. *Signs and Portents: Monstrous Births from the Middle Ages to the Enlightenment.* London: Routledge, 1991.

Wittgenstein, Ludwig. *Philosophical Investigations.* 1953. Reprint, London: Blackwell, 2001.

Wolfe, Cary. "Exposures." In *Philosophy and Animal Life,* ed. Stanley Cavell, Cora Diamond, John McDowell, Ian Hacking, and Cary Wolfe, 1–42. New York: Columbia University Press, 2008.

———. *What Is Posthumanism?* Minneapolis: University of Minnesota Press, 2010.

———, ed. *Zoontologies: The Question of the Animal.* Minneapolis: University of Minnesota Press, 2003.

World Bank. "The Sunken Billions: The Economic Justification for Fisheries Reform." http://siteresources.worldbank.org /EXTARD/Resources/336681-1224775570533/SunkenBillionsFinal.pdf (accessed 15 January 2011).

Yates, Julian. "Accidental Shakespeare." *Shakespeare Studies* 34 (2006): 90–122.

Youatt, William. *Cattle: Their Breeds, Management, and Diseases.* London: Baldwin and Cradock, 1834.

Žižek, Slavoj. *Interrogating the Real: Selected Writings.* Edited by Rex Butler and Scott Stephens. New York: Continuum, 2006.

———. *Looking Awry: An Introduction to Jacques Lacan Through Popular Culture.* Cambridge: MIT Press, 1992.

about the contributors

DEAN BAVINGTON is associate professor of geography at Memorial University of Newfoundland, St. John's, Newfoundland, Canada. He is the author of *Managed Annihilation: An Unnatural History of the Newfoundland Cod Collapse* (2010). His research addresses critical histories of environmental management focusing on fisheries and aquaculture.

RON BROGLIO is assistant professor in the Department of English, Arizona State University. In publications such as *Technologies of the Picturesque* (2008) and *Surface Encounters: Thinking with Animals and Art* (2011), his research focuses on how philosophy and aesthetics can help us rethink the relationship between humans and the environment.

MARK DION is an internationally exhibited visual artist who was born in New Bedford, Massachusetts, in 1961. He received a BFA (1986) and an honorary doctorate (2003) from the University of Hartford School of Art in Connecticut. Dion's work examines the ways in which dominant ideologies and public institutions shape our understanding of history, knowledge, and the natural world. He lives in Pennsylvania and New York City and works worldwide.

ERICA FUDGE is professor of English studies at the University of Strathclyde. She is the author of *Perceiving Animals: Humans and Beasts in Early Modern English Culture* (2000), *Animal* (2002), *Brutal Reasoning: Animals, Rationality, and Humanity in Early Modern England* (2006), and *Pets* (2008). She was the director of the British Animal Studies Network from 2007 to 2009 and associate editor for the humanities of the journal *Society and Animals* from 2002 to 2011.

JOAN B. LANDES is Walter L. and Helen Ferree Professor of Early Modern History and Women's Studies at Pennsylvania State University. Her books include *Women and the Public Sphere in the Age of the French Revolution* (1988), *Visualizing the Nation: Gender, Representation, and Revolution in Eighteenth-Century France* (2001), *Feminism: The Public and the Private* (1998), and *Monstrous Bodies / Political Monstrosities in Early Modern Europe* (2004). Her recent articles address designs for artificial life by eighteenth-century anatomists and makers of automata

PAULA YOUNG LEE researches the architecture of animal captivity. She is the editor of *Meat, Modernity, and the Rise of the Slaughterhouse* (2008) and the author of *Game: A Global History* (forthcoming) for Reaktion's Edible series. She has held fellowships from the University of Leipzig, the Institute for Humanities Research at Arizona State University, the Graham Foundation for Advanced Research in the Fine Arts, and the National Endowment for the Humanities. She is currently completing a book on the seventeenth-century menagerie at Versailles.

CECILIA NOVERO is a senior lecturer in the Department of Languages and Cultures at the University of Otago, New Zealand. Her book *The Antidiets of the Avant-Garde* (2010) and articles address the temporal relations between the historical avant-garde and neo-avant-garde, Dada, Viennese actionism, the Swiss artist Daniel Spoerri, the cultural history of food, and German and European film. She is an associate of the New Zealand Centre for Human-Animal Studies in Christchurch.

HARRIET RITVO's work on animals and history includes *The Animal Estate: The English and Other Creatures in the Victorian Age* (1987), *The Platypus and the Mermaid, and Other Figments of the Classifying Imagination* (1997), and *Noble Cows and Hybrid Zebras: Essays on Animals and History* (2010). She is the Arthur J. Conner Professor of History at the Massachusetts Institute of Technology.

NIGEL ROTHFELS is the author of a history of naturalistic displays in zoological gardens, *Savages and Beasts: The Birth of the Modern Zoo* (2002), and the editor of the collection *Representing Animals* (2002). He has received fellowships from Princeton University, the Australian National University, and the National Endowment for the Humanities, and is currently writing a history of ideas about elephants since the eighteenth century.

SAJAY SAMUEL is clinical professor at the Smeal School of Business, Pennsylvania State University. His research focuses on the history and politics of accounting and related calculative procedures. His current preoccupations center on the heterogeneity between common sense and techno-scientific rationality in the context of environmental management.

PIERRE SERNA is director of the Institut d'histoire de la Révolution Française and holds the chair of professor, history of the French Revolution, at the Université de Paris I, Panthéon Sorbonne. He is author of *Antonelle, aristocrate révolutionnaire, 1747–1817* (1997) and *La République des girouettes: Une anomalie politique: La France de l'extrême centre* (2005), and co-author of *Croiser le fer, culture et violence de l'épée dans la France moderne* (2007). He has recently begun working in the field of animals during the Revolution.

PAUL YOUNGQUIST teaches English at the University of Colorado. He writes on British romanticism, science fiction, and contemporary culture, and his books include *Monstrosities: Bodies and British Romanticism* (2003) and *Cyberfiction: After the Future* (2010).

index

PAGE NUMBERS IN *italics* INDICATE ILLUSTRATIONS.

a

o

p

r

s